I0148490

Submarine:
Hunter & Hunted

THE SURRENDER OF THE GERMAN SUBMARINE FLEET—
THE WHITE ENSIGN IS HOISTED OVER THE GERMAN EAGLE

Submarine:
Hunter & Hunted
British Submarine and Anti-Submarine
Operations During the First World War

Charles W. Domville-Fife

LEONAUR

Submarine: Hunter & Hunted—British Submarine and Anti-Submarine Operations During the First World War
by Charles W. Domville-Fife

Originally published in 1920 under the title
Submarine Warfare of To-day: How the Submarine Menace Was Met and Vanquished, With Descriptions of the Inventions and Devices Used, Fast Boats, Mystery Ships

Leonaur is an imprint of Oakpast Ltd

Material original to this edition and
presentation of text in this form
copyright © 2009 Oakpast Ltd

ISBN: 978-1-84677-976-3 (hardcover)
ISBN: 978-1-84677-975-6 (softcover)

http://www.leonaur.com

Publisher's Notes

The views expressed in this book are not necessarily those of the publisher.

Contents

Author's Note   7
Introduction   9
The Task of the Allied Navies   13
The New Navy—Training an Anti-Submarine Force   27
A Naval University in Time of War   35
The New Fleets in Being   37
The Hydrophone and the Depth Charge   54
Some Curious Weapons of Anti-Submarine Warfare   65
Mystery Ships   73
A Typical War Base   78
The Convoy System   90
The Mysteries of Submarine Hunting Explained   98
The Mysteries of German Mine-Laying Explained   111
The Mysteries of Minesweeping Explained   122
The Mine Barrage   140
Off to the Zones of War   147
A Memorable Christmas   151
The Derelict   158
Mined-In   163
The Casualty   171
How H.M. Trawler No. 6 Lost Her Refit   175
The Raider   180
The S.O.S.   184
In the Shadow of a Big Sea Fight   191
A Night Attack   198
Mysteries of the Great Sea Wastes   202
From Out the Clouds and Under-Seas   208
On the Sea Flank of the Allied Armies   218

Author's Note

I desire simply to say that I commenced taking an active interest in submarines in 1904. I wrote my first book on the subject, *Submarines of the World's Navies*, in 1910, and I have watched and written of the rise of these and kindred weapons for the past fifteen years of rapid development in peace and war, finally taking a humble part in the defeat of the great German submarine armada during the years 1914–1918.

C. D.-F.
1919

Introduction

While Great Britain remains an island, with dominion over palm and pine, it is to the sea that her four hundred millions of people must look for the key to all that has been achieved in the past and all that the future promises in the quickening dawn of a new era.

Not only over Great Britain alone, however, does the ocean cast its spell, for it is the free highway of the world, sailed by the ships of all nations, without other hindrances than those of stormy nature, and navigated without restriction from pole to pole by the seamen of all races. It was the international meeting-place, where ensigns were "dipped" in friendly greeting, and since the dawn of history there has been a freemasonry of the sea which knew no distinction of nation or creed.

When the call of humanity boomed across the dark, storm-tossed waters the answer came readily from beneath whatever flag the sound was heard. But in August, 1914, there came a change, so dramatic, so sudden, that maritime nations were stunned. Germany, in an excess of war fever, broke the sea laws, and laughed while women and children drowned. Crime followed crime, and the great voice of the Republican West protested in unison with that of the Imperial East. Still the Black Eagle laughed as it flew far and wide, carrying death to whomsoever came within its shadow, regardless of race and sex.

But there was an avenger upon the seas, one who had been rocked in its cradle from time immemorial, and to whom the world appealed to save the lives of their seamen. It sailed beneath the White Ensign and the Blue, and with aid from France, Italy and Japan it fought by day and by night, in winter gale and snow, and in summer heat and fog, in torrid zone and regions of perpetual ice to free the seas of the traitorous monster who had, in the twentieth century, hoisted the black flag of piracy and murder. For three years this ceaseless war was waged, and then, with her wonderful patience exhausted, the great sister nation of the mother tongue joined her fleets and armies with those of the battle-worn Allies and peace came to a long-suffering world.

In that abyss of war there was romance sufficient for many generations of novelists and historians. Many were the epic fights, unimportant in themselves, but which need only a Kingsley or a Stevenson to make them famous for all time. So with the happenings to be described in this book, many of them historically unimportant compared with the epoch-making events of which they formed a decimal part, but told in plain words; just records of romance on England's sea frontier in the years 1914-1918.

Although jealous of any encroachment on the space available for the description of guerrilla war at sea, there are many things which must first be said regarding the organisation and training of what may appropriately be termed the "New Navy," which took the sea to combat the submarine and the mine; also of the novel weapons devised amid the whirl of war for their use, protection and offensive power. Into this brief recital of the events leading to the real thing an endeavour will be made to infuse the life and local colour, which, however, would be more appropriate in a personal narrative than in a general description of anti-submarine warfare of to-day, but without which much that is essential could not be written without dire risk of tiring the reader before the first few chapters had been passed.

The names of places and ships have necessarily been changed to avoid anything of a personal character, and all references to existing or dead officers and men have been rigidly excluded as objectionable and unnecessary in a book dealing entirely with events.

Many of the incidents described—written while the events stood out in clear, mental perspective—could no doubt be du-

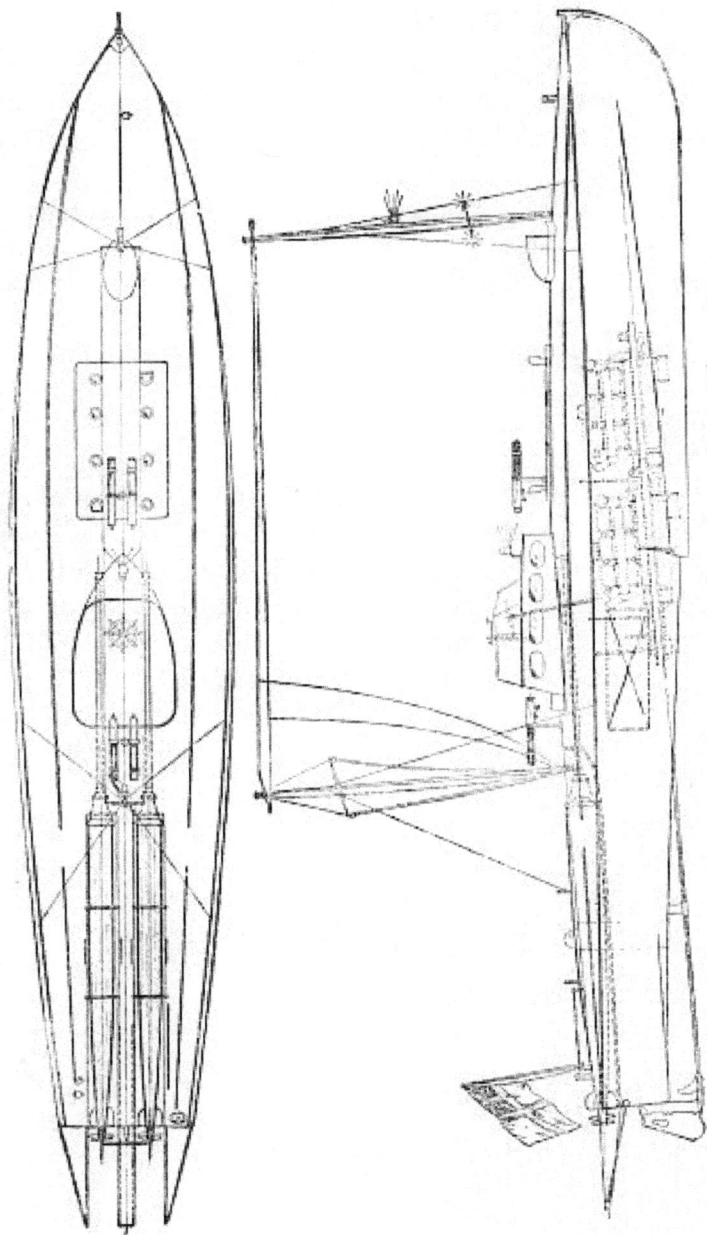

plicated and easily surpassed by many whose fortunes took them into zones of sea war during the historic years just past. If such is found to be the case, then the object of this book has been accomplished, for it sets out to tell, not of great epoch-making events, but of the organisation, men, ships, weapons and ordinary incidents of life in what, for lack of a better term, has been called the "New Navy"—a production of the World War.

It may be that an apology is due for placing yet another war book before a war-weary public, but an effort has been made to make of the following chapters *a record of British maritime achievement*, more than a narrative of sea fighting, although to do this without introducing the human element, the arduous nature of the work, the monotony, the danger and, finally, the compensating moments of excitement would have been to falsify the account and belittle the achievement.

There are many books available, full of exciting stories of sea and land war, but no other, so far as the Author knows, which describes in detail and in plain phraseology those important "little things"—liable to be overlooked amid the whirl of war—which go to make an anti-submarine personnel, fleet and base, together with an account of "how it was done."

Chapter 1

The Task of the Allied Navies

The hour was that of the Allies' greatest need—the last months of the year 1914. On that fateful 4th August the British navy was concentrated in the North Sea, and the chance for a surprise attack by the German fleet, or an invasion of England by the Kaiser's armies, vanished for ever, and with this one chance went also all reasonable possibility of a crushing German victory.

Although during the years of bitter warfare which followed this silent *coup de main* the German fleet many times showed signs of awakening ambition, it did not, after Jutland, dare to thrust even its vanguard far into the open sea. Behind its forts, mines and submarines it waited, growing weaker with the dry-rot of inaction, for the chance that fickle Fortune might place a single unit of the Allied fleet within easy reach of its whole mailed-fist.

With a great and modern fleet—the second strongest in the world—awaiting its chance less than twenty hours' steam from the coast of Great Britain, it quickly became evident that the old Mistress of the Seas would have to call upon her islanders to supply a "new navy" to scour the oceans while her main battle squadrons waited and watched for the second Trafalgar.

Faced, then, with the problem of a long blockade, a powerful fleet in readiness to strike at any weak or unduly exposed point of land or squadron, and with similar problems on a decreasing scale imposed by Austria in the Adriatic and by Turkey behind the Dardanelles, the work of the main battle fleets became well defined by the commonest laws of naval strategy.

All this without taking into account the widespread menace

of submarines and mines, and, in the earlier stages of the war, the rounding-up of detached enemy squadrons, such as that under Von Spee in South American waters, and the protection of the transport and food ships from raiders like the *Wolfe* and the *Moewe*.

The German High Command realised this as quickly as that of the Allies. Their overseas commerce was strangled within a few days of the Declaration of War with Great Britain, and their fleet was confined to harbour, with the exception of occasional operations against Russia in the Baltic. From the German standpoint the naval problem resolved itself into one of how best to strike at the lines of communication of the Allies, paying special attention, first, to the transport of troops, and, second, to England's food supply. As they alone knew to what extent they would violate the laws of war and of humanity, it became apparent that the submarine and the mine were the only possible weapons which could be used for this purpose in face of the superior fleets of the Allies. But the number of these weapons was strictly limited compared with the immense shipping resources at the command of the Western Powers, so one submarine must do the work of many, and an effort was made to accomplish this by a reign of sea terrorism and inhuman conduct unparalleled in the history of the world. It opened with the sinking of the *Lusitania*.

The Allies had secured and maintained the command of the sea, and *all that it implies*, but to do this with the certainty of correct strategy they had to dedicate almost their entire battle fleet to the purpose for which battle fleets have always been intended—the checkmating or annihilation of the opposing navy.

There came a second problem, however, one entirely new to sea warfare, and unconsidered or provided against in its strategic and tactical entirety because hitherto deemed too inhuman for modern war. This was the ruthless use of armed submarines against unarmed passenger and merchant ships, and the scattering broadcast over the seas, regardless of the lives and property of neutrals, of thousands of explosive mines.

The type of ship constructed exclusively for open sea warfare against surface adversaries was not the best answer to the submarine. The blockading of the hostile surface fleet did not prevent, or even greatly hinder, the free passage of submarine flotillas, and the building by Germany of under-water mine-layers enabled fields of these weapons to be laid anywhere within the carrier's radius of action.

In this way the second, or submarine, phase of the naval war opened, and it was to supplement the comparatively few fast destroyers and other suitable ships which could be spared from the main fleets that the "new navy" was formed.

THE SHIPS

The area of the North Sea alone exceeds 140,000 square miles, and when the whole vast stretch of water encompassed by what was known as the radius of action of hostile submarines, from their bases on the German, Belgian, Austrian, Turkish and Bulgarian coasts, had to be considered as a possible zone of operations for German and Austrian under-water flotillas, much of the water surface of the world was included. Likewise the network of sea communications on which the Allies depended for the maintenance of essential transport and communication comprised the pathways of the seven seas. To patrol all these routes adequately, and to guard the food and troop ships, hastening in large numbers to the aid of the Motherland from the most distant corners of the earth; to protect the 1500 miles sea frontier of the British Isles; to give timely aid to sinking or hard-pressed units of the mercantile fleet; to hound the submarine from the under-seas and to sweep clear, almost weekly, several thousand square miles of sea, from Belle Isle to Cape Town and the Orkneys to Colombo, required ships, not in tens, but in thousands. To find these in an incredibly short space of time became the primary naval need of the moment.

Who that lived through those days will forget the struggle to supply ships and guns? The searching of every harbour for craft, from motor boats to old-time sailing-ships, and from fishing craft to liners. The scouring of the Dominions and Colonies. How blessed was their aid! Help, generous and spontaneous, came from all quarters, including the most unexpected. Over five hundred fast patrol boats, or motor launches, in less than twelve months from Canada and America. Guns from Japan. Coasting steamers from India, Australia, New Zealand and South Africa. Seaplanes from the Crown Colonies. Rifles from Canada. Machine guns from the United States. Ambulances from English and Colonial women's leagues. In fact, contributions to the "new navy" from all corners of the earth.

To patrol the coasts of Britain alone, and to keep its harbours and coastal trade routes clear of mines, needed over 3500 ships, with at least an equal number of guns, 30,000 rifles and revolvers, and millions of shells.

In addition to this huge fleet other smaller squadrons were required for the Mediterranean, the Suez Canal and Red Sea, the East and West Indies, the coasts of the Dominions and Colonies, and for the Russian lines of communication in the White Sea. For these overseas bases just under 1000 ships were required, exclusive of those locally supplied by the Dominions and Colonies themselves.

All this without considering the main battle fleets or, in fact, any portion of the regular navy, and the ships required for the transport of food, troops and munitions of war, together with their escorts. Some idea of the numbers engaged in keeping the Allies supplied with the diverse necessities of life and war may be gathered from the fact that the average sailings in and out of the harbours of the United Kingdom alone during the four years of war amounted to over 1200 a week.

The immense fleet forming the new navy was not homogeneous in design, power, appearance or, in fact, in anything except the spirit of the personnel and the flag beneath which they fought—and alas! nearly 4000 died. The squadrons, or units, as they were called, consisted of fine steam yachts, liners from the ocean trade routes, sturdy sea tramps, deep-sea trawlers, oilers, colliers, drifters, paddle steamers, and the more uniform and specially built fighting sloops, whalers, motor launches and coastal motor boats. The latter type of craft was aided by its great speed, nearly fifty miles an hour; but more about these ships and their curious armament later.

The great auxiliary navy had to be built or obtained without depleting the ordinary mercantile fleets, and the shipbuilding and repairing yards, even in the smallest sea and river ports, worked day and night. The triumph was as wonderful as it was speedy. In less than fifteen months from August, 1914, the new navy was a gigantic force, and its operations extended from the Arctic Sea to the Equator. All units were armed, manned and linked up by wireless and a common cause. Before this could be accomplished, however, the problem of maintaining this vast fleet and adequately controlling its operations had to be faced and overcome. The seas adjacent to the coasts of the United Kingdom, the Mediterranean Littoral and Colonial waters were divided into "patrol areas" on special secret charts, and each "area" had its own naval base, with harbour, stores, repairing and docking facilities, intelligence centre, wireless and signal stations, reserve of officers and men, social headquarters, workshops and medical department.

Each base was under the command of an admiral and staff, many of the former returning to duty, after several years of well-earned rest, as captains and commodores, with salaries commensurate with their reduced rank. Their staffs consisted of some six to twelve officers of the new navy, with possibly one or two from the "*pukka* service," and their command often extended over many hundreds of square miles of submarine and mine infested sea. Of these bases, which will be fully described in later chapters, there were about fifty, excluding the great dockyards and fleet headquarters, but inclusive of those situated overseas. When it is considered what a war base needs to make it an efficient rendezvous for some hundreds of ships and thousands of men, some idea of the gigantic task of organisation which their establishment, often in poorly equipped harbours and distant islands, required, not only in the first instance, but also with regard to maintenance and supplies, will be realised, perhaps, however, more fully when it is stated that the average ship needs a month spent in docking and overhauling at least once a year, and that the delicate and more speedy units of such a fleet need nearly four times that amount of attention.

One of the first requisites of the auxiliary navy was the creation of a headquarters staff at the Admiralty, London. This was formed from naval officers of experience both in the regular service and in the two reserves (R.N.R. and R.N.V.R.). Forming an integral part of the great British or Allied armada, all operations were under the control of the Naval War Staff, but for purposes of more detailed organisation and administration additional departments were created which exercised direct jurisdiction over their respective fleets. The principal of these was known as the "Auxiliary Patrol Office," under the Fourth Sea Lord and the Department of the Director of Minesweeping. These formed a part of the General Staff—if a military term is permissible—and both issued official publications periodically throughout the war, which served to keep the staffs of all the different war bases and the commanding officers of the thousands of ships informed as to current movements and ruses of the enemy. It is unnecessary to detail more closely the work of these departments, especially as much has yet to be said before plunging into the maelstrom of war. A sufficient indication of the colossal nature of the work they were called upon to perform will be found in a moment's reflection of what the administration and control of such a large and nondescript fleet, spread over the world—from the White Sea to the East Indies—must have meant to the small staff allowed by the exigencies of an unparalleled war.

OFFICERS AND MEN

The greatest problem in modern naval war is, undoubtedly, the supply of trained men. For this reason it has been left to the last to describe how the difficulty was faced and overcome by England and her overseas Dominions in 1914. Before doing so, however, it may be of interest to give here a few extracts from an excellent little official publication, showing how the British fleet was manned and expanded in bygone days of national peril:[1]

---

1. Extract from *Naval Demobilisation*—issued by the Ministry of Reconstruction.

In time of war there has always been an intimate connection between the Royal Navy and the Merchant Service. Latterly, and more especially since the Russian War of 1854 to 1856, this fact tended to be forgotten, partly because men-of-war developed on particular lines and became far more unlike merchantmen than they had ever been before, and also because, by the introduction of continuous service, the personnel of the Navy seemed to have developed into a separate caste, distinguished by its associations, traditions and *esprit de corps*, as much by its special training and qualifications, from other seafaring men. This war has proved once again, to such as needed proof, that the two services cannot exist without each other, and that the Sea Power of the Empire is not its naval strength alone, but its maritime strength. Even at the risk of insisting on the obvious, it is necessary to repeat that, for an Island Empire, a war at sea cannot be won merely by the naval action which defeats the enemy; naval successes are of value for the fruit they bear, the chief of which is the power that they give to the victor to maintain his own sea-borne trade and to interrupt that of the enemy.

An elementary way of looking at the problems of manning the Royal Navy and the Merchant Service is to consider that there is in the country a common stock of seamen, on which both can draw. But this theory, like many others equally obvious and tempting, has the disadvantage that it leaves important factors out of account and, if worked out, results in an absurdity. Thus, shortly before war began there were in the country some 420,000 seamen, of whom one-third were in the Navy and two-thirds engaged in merchant ships and fishing vessels. There was no considerable body of unemployed seamen. During the war the personnel of the Navy was expanded to something like the 420,000 which represents the common stock of seamen. Therefore, if the theory met the case, there would have been no men left for the Merchant Service. But the merchant ships, in spite of difficulty and danger, continued to run, employing great numbers of men. And we must not forget to take into account the number of men, amounting to 48,000 killed and 4500 prisoners of war, who have been lost in the two services during the war. So it comes to this, that the common stock of seamen, or at least of men fit to man ships, has expanded during the war by more than 50 per cent. Whence came these extra men? Clearly for the most part from the non-seafaring classes.

The Navy in November, 1918, employed some 80,000 officers and men from the Merchant Service—*viz.* 20,000 R.N.R. ratings, 36,000 Trawler Reserve, and 20,000 mercantile seamen and firemen on Transport agreements, plus the officers. If the supposition, made in the absence of statistics, is correct that at this time the number of men in the Merchant Service itself had decreased proportionately to the loss of tonnage, it would seem that the Merchant Service needed no considerable inflow of men during the war. In other words, most of those added to the stock of seamen during the war must have gone into the Navy. This corresponds with known fact: the Navy has, in addition to the Reserve men already mentioned, nearly 200,000 men to demobilise in order to put its personnel on the footing on which it stood when war broke out.

It will be of interest to see how the personnel of the Navy expanded in former wars, and how at the peace it was invariably reduced to something like its pre-war figures. This can readily be done in tabular form:

Naval Personnel (Numbers Voted)

Year	War	Before the War	Maximum during War	After the Peace
1689	League of Augsberg	7,040	—	—
1697		—	40,000	—
1700				7,000
1700	Spanish Succession	7,000	—	—
1712		—	40,000	—
1713		—	—	10,000

Year	War			
1738	Austrian Succession	10,000	—	—
1748		—	40,000	—
1759		—	—	10,000
1754	Seven Years' War	10,000	—	—
1762		—	70,000	—
1764		—	—	16,000
1775	American	18,000	—	—
1783	Independence	—	110,000	—
1785		—	—	18,000
1793	French Revolution	16,000	—	
1801		—	135,000	
1803		—	—	50,000
1803	Napoleonic War	50,000	—	—
1812		—	145,000	—
1817		—	—	19,000
1853	Russian War	45,500	—	—
1856		—	76,000	—
1857		—	—	53,000
1914	The Present War	146,000	—	—
1918		—	450,000	—

It appears at once from these figures that the naval expansion during earlier wars was in most cases much greater proportionately than it has been in this. Roughly the personnel in this war has been multiplied by three; in earlier wars it was increased six, seven, eight, or even nine fold, if we take the difference between the figures for 1792 and 1812.

It is a common error to suppose that our ships in the old wars were manned entirely by seamen. A knowledge of how the men were raised shows that this cannot have been so; and confirmation can be had from a very brief study of ships' muster books. Only about a third of the crew of a line-of-battle ship were, in the seaman's phrase, 'prime seamen.' The rest were either only partly trained or were frankly not sailor men. The *Victory* at Trafalgar was not an ill-manned ship—here is an analysis of her crew: officers, commissioned and warrant, 28; petty officers, including marines, 63; able seamen, 213; ordinary seamen and boys, 225; landsmen, 86; marines, 137; artificers, 18; quarter gunners, 12; supernumeraries and domestics, 37.

During the whole of our naval history down to 1815 it was the invariable rule that in peace time the battle fleets were laid up unmanned, and only enough ships were kept in commission to 'show the flag' and to police the sea. This accounts for the very large increase of the naval personnel which immediately became necessary when there was a threat of war; and it accounts also for the difficulty which was always experienced in raising the men. This difficulty was even greater than we are apt to suppose, for the Merchant Service has never been able to give the navy more than a fraction of the total number of men needed, and the machinery for raising extra men has, until this war, always been of a most primitive nature.

When war came the ships were commissioned, without crews. This could be done because from the latter part of the seventeenth century there was a permanent force of officers. Then the officers had to find their own crews. They began by drawing their proportion of marines, and then proceeded to invite seamen to volunteer. In this way they got a number of skilled seamen, men who had been in the navy before, and came back to it either as petty officers or in the hope of becoming so. Then warrants to impress seamen would be issued. Theoretically the impress was merely a form of conscription, the Crown claiming by prerogative the right to the services of its seafaring subjects. Practically a good deal of violence was at times necessary, as many of the men, preferring to sail in merchant ships, or wishing to wait for a proclamation of bounty, tried to avoid arrest. The scuffles that took place on these occasions gave the impress service a bad name, not altogether deserved, for real efforts were made to avoid hardship, and in any case the number of men raised in this way was greatly exaggerated by popular report.

There was no compulsion during the Great War to join any unit of the British fleet. Therefore all were either in the regular service, reservists or volunteers. The need was made known not only throughout the British Isles, but also from Vancouver to

Cape Town, Sydney and Wellington, and men in all walks of life, but with either the *Wander-Lust* or true love of the wide open sea in their blood, rallied from all parts of the far-flung Empire to the call of the White Ensign.

In order to obtain some 6000 officers and nearly 200,000 trained or semi-trained men, new sources of supply had to be tapped. Already the great battle fleets, brought up to full war strength and with adequate reserves, had absorbed nearly all the Reserve officers who could be spared from the food and troop transports.[2]

First came the great sea-training establishment of the Empire—the Mercantile Marine and its retired officers and men—already heavily depleted. Then the yacht clubs from the Fraser to the Thames and Clyde. Thousands of professionals and amateurs came overseas to the training cruisers and the "naval university," Canada alone supplying several hundred officers.

Doctors came from the hospitals and from lucrative private practices. The engineering professions and trades supplied the technical staffs and skilled mechanics. The great banks and city offices yielded the accountants, and the fishing and pleasure-boating communities, not only of Great Britain, but also of the Dominions and Colonies, yielded the men in tens of thousands. In this way the personnel of the new navy was completed in a very few months.

Before passing on to describe, in the detail of personal acquaintance, the severe training of this naval force, a general knowledge of its heterogeneous character is necessary to enable the reader to understand this great assemblage of the sons of the Empire.

In the smoke-filled wardroom and gunroom of the training cruiser, H.M.S. *Hermione* one windy March evening in 1916 there were some eighty officers of the auxiliary fleet, and of this number one hailed from distant Rhodesia, where he was the owner of thousands of acres of land and a goodly herd of cattle, but who, some time in the past, had rounded the Horn in a

---

2. The personnel of the new navy consisted of R.N., R.N.R. and R.N.V.R. officers. The former came mostly from the retired list. The R.N.R. needed training only in such subjects as gunnery, tactics, etc. The training of the R.N.V.R. is here described.

A LARGE AND HEAVILY ARMED GERMAN SUBMARINE OF THE CRUISER TYPE

wind-jammer and taken *sights* in the "Roaring Forties." Another was a seascape painter of renown both in England and the United States. A third was a member of a Pacific coast yacht club. A fourth was the son of an Irish peer, the owner of a steam yacht. Then came a London journalist, a barrister, a solicitor and a New Zealand yachtsman, while sitting at the table was a famous traveller and a *pukka* commander.

In the neighbouring gunroom, among the crowd of sub-lieutenants—all of the same great force, the Royal Naval Volunteer Reserve—was a grey-haired veteran from the Canadian Lakes, a youngster from the Clyde, the son of a ship-owner from Australia and a bronzed mine manager from the Witwatersrand.

Among the engineers and mechanics the same diversity. Men from several of the great engineering establishments, a student from a North Country university, electrical engineers from the power stations and mechanics from the bench, with here and there one or two with sea-going experience.

In the forecastle and elsewhere about the old cruiser—now merely a training establishment—were sailors with years of experience in both sail and steam. Fishermen from the Hebrides and Newfoundland rubbing shoulders with yacht hands from the Solent and Clyde.

From this curiously mixed but excellent raw material a naval personnel, with its essential knowledge and discipline, had to be fashioned in record time by an incredibly small staff of commissioned and warrant officers of the permanent service, aided by the more experienced amateurs.

It must, however, not be thought from this that the amateur was converted into a professional seaman in the space of a week or two. Three months of specialised training enabled them to take their place in the new fleet, but with some it required a much longer period to enable them to feel that perfect self-confidence when *alone* in the face of difficulties and dangers which is the true heritage of the sea.

To describe here the training of officers and men would be to repeat what will be more fully and personally described in succeeding chapters. It is sufficient to say that the aim was to bring them all to a predetermined standard of efficiency, which would enable the officers to command ships of specific types at sea and in action, and the men to form efficient engineers and deck hands for almost any ship in the Navy.

The medical branches naturally required no special training and the accountants merely a knowledge of naval systems of financial and general administration. These two branches had their own training establishments.

When the period of preliminary training in the cruiser *Hermione* was over the officers were passed on to the Royal Naval College at Greenwich, and from there to one or other of the fifty war bases in the United Kingdom, the Mediterranean or farther afield. Their appointments were to ships forming the fleets attached to each of these bases and generally operating in the surrounding seas.

In this way the whole zone of war was covered by an anti-submarine and minesweeping organisation and general naval patrol, which operated in conjunction with, but separate from, the battle fleets, squadrons and flotillas, which were thus left free to perform their true functions in big naval engagements.

CHAPTER 2

The New Navy—
Training an Anti-Submarine Force

Having described the *raison d'être* of the new navy, and how it became a fleet in being, with its own admirals, captains, staffs, bases and all the paraphernalia of war, I can pass on to a more intimate description of the training of the officers and men, preparatory to their being drafted to the scattered units of this great anti-submarine force.

Lying in the spacious docks at Southampton was the old 4000-ton cruiser *Hermione*, which had been brought round from her natural base in Portsmouth dockyard to act as the depot ship and training establishment for a large section of this new force. Not all the officers and men of the auxiliary fleet were, however, destined to pass across its decks. This vessel was reserved for the Royal Naval Volunteer Reserve, from which a very considerable proportion of the entire personnel of the new fleet was drawn. Nor was H.M.S. *Hermione* the first depot ship of the war-time R.N.V.R. at Southampton, for the Admiralty yacht *Resource II.* had been used for the first few drafts, but was unfortunately burned to the water's edge. There were also other vessels and establishments at Portsmouth, Devonport and Chatham. These were, however, mainly for the reception and brief training of the more experienced Merchant Service officers, entered in the Royal Naval Reserve for the duration of the war, and for the surgeons and accountants.

The men of the new force were mostly trained in the naval barracks and depot ships situated at the big naval centres, such as Portsmouth and Chatham. After a few weeks all these establishments were drafting, in a constant stream, the trained human element to the vessels awaiting full complements at the different war bases, or being constructed in the hundreds of shipyards of the Empire.

About H.M.S. *Hermione*, which has been selected as being representative of the training depots of a large section of the auxiliary service, little need be said, beyond the fact that she was commanded, first, by a distinguished officer from the Dardanelles, and subsequently by an equally capable officer, who, by the irony of fate, had in pre-war times been a member of the British Naval Mission to the Turkish navy—both of them men whose experience and unfailing tact contributed largely to the success of the thousands of embryo officers trained under their command.

The ship herself was a rambling old cruiser, but very little of the actual training was carried out on board. Spacious buildings on the quayside provided the training grounds for gunnery, drill, signalling, engineering and all the complicated curricula, of which more anon. Lying in the still waters of the dock, alongside the comparatively big grey cruiser, were the trim little hulls of a numerous flotilla of 20-knot motor launches, newly arrived from Canada, with wicked-looking 13-pounder high-angle guns, stumpy torpedo-boat masts and brand-new White Ensigns and brass-bound decks. These were the advance guard of a fleet of over 500 similar craft, to the command of which many of the officers being trained would, after a period of practical experience at sea, eventually succeed.

There were besides numerous other mosquito craft, which throbbed in and out of the dock from that vast sheltered arm of the sea called Southampton Water on mysterious errands, soon to be solved by new recruits in the chilly winds of winter nights and early mornings.

This, then, was the mother ship and her children. When once the aft gangway leading up from the dockside to the clean-scrubbed decks had been crossed, and the sentry's challenge answered, the embryo officer left civilian life behind and commenced his training for the stern work of war.

It may not be out of place to give here a closer description of the training of the officers and men of the new navy, drawn from personal experience. To do this without the irritating egoism of the personal narrative it will be necessary, as often in future pages, to adopt the convenient "third person."

****

The night was fine, but a keen March wind blew from off the sea. The dock lights were reflected in the still waters of the harbour. Tall cranes stood out black and clearly defined against the cold night sky. The shadows were deep around the warehouses, stores and other buildings of the busy dockside.

Lying in the south-western basin was the big grey hull of the cruiser, newly painted, and looking very formidable, with its tall masts and fighting-tops towering into the blue void, and its massive bow rising high above the dock wall.

Coming from the darkness on board were the tinkling notes of a banjo and the subdued hum of voices. Then the loud call of the quartermaster and the ringing of eight bells.

A group of newly appointed officers picked their way carefully among the tangled mooring ropes on the quayside and as they approached the warship were duly challenged by the sentries. Two of them had only just arrived from distant New Zealand. They were all "for training," and on mounting the quarterdeck gangway were politely requested by the smiling quartermaster to report at the ship's office. In order to get from the deck to this abode of paymasters and writers, except by the tabooed "captain's hatchway," there had to be negotiated a long passage leading past the wardroom and the gunroom. In normal times at such an hour this passage would probably have been almost deserted, with the exception of a sentry, but the training was being speeded up to meet the demands of war, and with nearly 200 officers, many of whom fortunately lived ashore, constantly moving to and fro, it became either a semi-dark, congested thoroughfare, in which everyone was curtly apologising for knocking against someone else, or else it contained the steady pressure of a gunroom overflow meeting, with a tobacco-scented atmosphere peculiarly its own.

When the formality of reporting arrival had been completed, the embryo officers were taken in tow by the "Officer of the Day," whose duty it was to introduce them to the gunroom and make them familiar in a general way with the routine of the ship. The officer who performed this ceremony on the night in question has since held a highly responsible post at the Admiralty—such is the fortune of war.

The first shock came when the work for the following day was explained. It commenced with physical drill on the quayside at 7 a.m. and ended with instruction in signalling at 6 p.m.!

****

The early morning was bitterly cold but fine. Physical "jerks" was not a dress parade; in fact, some of the early risers on the surrounding transports and ocean mail boats must have wondered what particular form of mania the crowd of running, leaping and arm-swinging men, in all stages of undress on the quayside, really suffered from.

Breakfast and Divisions were the next items on the programme, and the new-comers looked forward to the day's work with the keen interest of freshness.

Morning Divisions and Evening Quarters are events of some importance in the daily routine of his Majesty's ships. They are parades of the entire ship's company, with the exception of those on important duty, marking the beginning and end of the day's work. The crew, or men under training, are mustered in "Watches," under their respective officers, and stand to attention at the bugle call. The senior officer taking divisions then enters, a roll is called and the names of those absent reported. The chaplain stands between the lines of men; the order "Off caps!" is given and prayers commence. When these are finished certain orders of the day are read out to the assembled ship's company and the parade is over.

At evening quarters, on certain days in the week, the names were read out of the officers and men detailed for special duties or for draft to a zone of war.

When morning divisions were over the day's work began.

The embryo officers were attached to the seamanship class, consisting of about twenty men of all ages. Oilskins were donned, for the sky was overcast and the wind keen. They climbed down the steel sides of the cruiser on to the small deck of a tender, which was to convey them out on to the broad but sheltered waters where much of the preliminary practical training was to take place during the following weeks.

The instructor, an officer attached for the purpose, then divided his class into two "watches," one being directed to work out the proposed course of the ship on the charts in the cabin and to give the necessary orders to the other watch on deck, who were to carry them into effect as the ship steamed along, with the aid of sextant, compass, wheel, engine-room telegraph, lead and log-line. As all possessed some knowledge of the sea, and had experience in navigating, this work did not prove as difficult as it undoubtedly would to anyone entirely devoid of nautical knowledge.

Those in the cabin with the charts worked out the compass courses from one point to another, making the necessary allowances for tide, deviation, etc. Others of the same watch received reports from the "bridge" and made the correct entries in the log-book. All elementary work, but which needed practice to make perfect, and on the accuracy of which men's lives would depend in the very near future.

The watch on deck was engaged in the more practical work of coastal navigation and could see the effect of any mistake made theoretically by their companions below. At midday the watches were reversed. Those working at the charts and courses came on deck and the seamen of the morning became the navigating officers of the afternoon.

On this particular day the second or port watch had the worst of it. A squally wind and rain had set in, making the work on deck thoroughly wet and uncomfortable. An hour or so later the small ship was rolling and pitching and everyone was drenched. The lead was kept going by hands numb with cold—a foretaste of the long and bitter days and nights to be afterwards spent in wintry seas.

The training cruises were continued for many days and were interspersed with lectures on the elements of good seamanship, the more advanced theory and practice of navigation being left for a later course at the Royal Naval College, Greenwich.

After seamanship came gunnery. Each of the different types of heavy but finely made weapons had to be learned in detail—a feat of memory when it came to the watch-like mechanism of the Maxim. Guns were disabled and had to be put right. They missed fire and were made by the instructors—old naval gunners—to play every dastardly trick conceivable. The final test which had to be successfully passed was the dismantling of each type of gun used in the auxiliary fleet and the reassembling of it.

With gunnery came also the marks and uses of the different kinds of ammunition, the systems of "spotting" and "range-finding." Every gun had its officer crew and the rapidity of fire was recorded. Each man in turn was chosen to give the necessary orders and to judge the ranges and deflections. In this way not only was the practical work learned by heart, but also the theory of naval gunnery, so far as it related to the smaller types of weapon.

The use of the depth charge, both mechanically and tactically, was expounded and practically demonstrated, together with that of the torpedo, the mine, mine laying and sweeping, and the peculiarities of various explosives. Rifle and revolver practice was encouraged, and Morse and semaphore signalling formed part of the daily routine.

The training was not entirely preparatory for work afloat. Squad and company drill, rifle and bayonet exercise, and manoeuvring in extended order formed a part of the comprehensive training. One day, not many weeks after their arrival, the officers whose fortunes have been followed found themselves shouting orders and directing by arm and whistle lines of dusty *camarades* advancing over a common in the most approved military fashion.

The training was not all hard work. The gathering of so many men from all quarters of the world, with a wealth of experience and adventure behind them, was in itself a source of mutual interest—and incidentally an education in modern British Imperialism. Scarcely any part of the world went for long unrepresented in either the wardroom or gunroom of the old cruiser *Hermione* in those days of war, and many were the yarns told of Alaska days, hunting in Africa, experiences in remote corners of North America, pearling in the Pacific and life on the Indian frontier, to say nothing of wild nights on the seven seas. Grey heads and round, boyish faces, the university and the frontier, with a camaraderie seldom equalled.

The period of training in the old cruiser was drawing to a close when each officer was appointed to "Boat Duty." There were five launches on duty at a time, and their crews had to be instantly ready day and night. The most coveted were the two 21-knot boats, used almost exclusively for the conveyance of pilots to and from the hospital ships and transports. Then came the patrol boat, a slow old tub with a comfortable cabin, and work out on Southampton Water at night. The three "duty boats" were for emergency use and were held at the disposal of the naval transport officer.

The duties on each boat varied and were in the nature of training. The pilot boat was required to lie alongside the cutter, out beyond the harbour, and to convey the pilots at high speed to and from the stream of shipping. It was a pleasant duty which entailed alternate nights in the generous, breezy company of the old sea-dogs of the cutter, with occasional races at half-a-mile a minute through the darkness and spray to the moving leviathans of the ocean.

The patrol ambled up and down the sheltered waterways during the day and night, examining the "permits" of fishermen and preventing the movement of small craft during the hours of darkness, when the long lines of troop-ships were leaving for France.

The work of the duty boats varied from day to day, but there was always the morning and evening mail to be collected from and delivered to the ships of the auxiliary fleet lying out in the fair-way.

When this spell of water-police work was over there came

a few days' practice in the handling of the fast sea-going patrol launches, or "M.L.'s," about which so much has since been written in the daily papers.

After the cramming received in the lecture-rooms, the arduous drill and the somewhat monotonous work on the slow-moving tenders, the runs seaward on these new and trim little vessels, the manoeuvring at nineteen knots, the breeze of passage and the feeling of controlled power acted as an elixir on both mind and body. Then came firing practice in the open sea. The sharp crack of cordite, the tongues of livid flame, the scream of the shells, the white splashes of the ricochet and the salt sea breezes.

Two days later the preliminary training was over and there loomed ahead a period of hard study at the Royal Naval College.

Chapter 3

A Naval University in Time of War

Built by King Charles I. for the Stuart navy, and used for over two and a half centuries as the university of the Senior Service, the Royal Naval College, Greenwich, is a building with an historic past. It has housed, fed and taught many of England's most illustrious sailors.

It was to cabin and lecture hall in this fine old building that officers of the new navy went to complete their knowledge of navigation and kindred subjects when their preliminary sea training came to a close.

There is but little romance in a highly specialised course of study designed to enable the recipients to find their way with safety, both in sunshine and storm, over the vast water surface of the world. To describe here the subjects taught would only be wearisome and uninteresting. Sufficient to say that the course was a most comprehensive one and admirably arranged by masters of the mariner's art. If any fault can be found it is certainly not one of paucity of information, and the proof of its efficacy can be found in the fact that, so far as the author knows, there was not a single ship, afterwards commanded by officers who underwent this training, lost through insufficient knowledge of the art of navigation.

The days spent in the Naval College were fully occupied by attendance at lectures and the evenings in private study and the preparation of elaborate notes and sketches for the final passing-out examination. There was one moment of each day which was rendered historic by old custom. It came at the conclusion of dinner in the big white hall, when the officer whose turn it happened to be rose to his feet and gave the toast of the navy— "Gentlemen, the King!"

It was in the grounds of this college that many officers saw their first zeppelin raid. On one occasion it occurred late in the fourth week of the course. Nearly all were in their respective studies, surrounded by a mass of papers, charts, drawing instruments and books, making the last determined attack on various knotty problems previous to the final examination.

Ten p.m. had just been registered by the electric clocks in the famous observatory overlooking the college, when the sound of running feet came down the long corridors. A stentorian voice shouted: "All lights out!"

In a moment the whole building, with its labyrinth of corridors, was plunged into Ethiopian darkness. Doors were opened and a jostling crowd of men groped their way down passages and stone staircases into the grounds. Here the Admiral and his staff were making sure that no lights were visible. Traffic in the near-by thoroughfare had been stopped, and all around lay the Great Metropolis, oppressively dark and still.

A searchlight flashed heavenwards and was followed by other beams. All of these suddenly concentrated on the gleaming white hull of a zeppelin, high in the indigo sky. The ground trembled under the fire of the anti-aircraft batteries. Shells whistled and moaned over the College and bright flashes came from little puffs of white smoke high in the central blue.

Dull-sounding but earth-shaking booms came from different points as the airship dropped her deadly cargo. Shrapnel fell on the congested house-tops with a peculiar hiss and thud and ambulances rumbled over the stone-paved high-road.

It was a small incident and scarcely worth the space required for its recording, but it served a purpose—to steel the heart and steady the hand for the time to come.

CHAPTER 4

The New Fleets in Being

Back once again on the old cruiser with training completed and awaiting draft to the zones of war. Then came the sailing orders. The name of each officer was called in turn and he disappeared into the ship's office, to return a few minutes later carrying a sheaf of white and blue Admiralty orders, his face grave or gay according to destination.

Some were for the Spanish Main and bemoaned their fate at being ordered to a station so remote from the principal zone of war. Others were destined for the Mediterranean and comforted themselves with hopes that trouble was brewing elsewhere than in the Adriatic, to which a lucky few were appointed. The Suez Canal and Egypt claimed their share, but by far the greater number were bound for the misty northern seas.

About the training given to the 200,000 men little can be said here because of its diversity. They came as volunteers from all quarters of the globe, were collected at the great depots in Portsmouth, Chatham and Devonport, were trained in the art of signalling, squad drill, gunnery, seamanship and the hundred and one things required by the "handy man," then belched forth into the ships.

Some had sailed the sea for years before in vessels of all kinds and needed little more than the sense of cohesion and unquestioning obedience imparted by discipline and drill. Others knew more of the working of a loom, or the extraction of coal, than of seamanship, and spent a cheerful but arduous few months in training depots and on special ships completing their education.

Cooks there were who could make little else besides Scotch broth, while others, the engineers—or motor mechanics, as they were called when appointed to some of the petrol-driven patrol boats—knew their profession or trade better than they could be taught, and proved themselves untiring and indomitable when it came to the real thing—as will be seen later.

Having now described the training of both officers and men, we come to the ships they were called upon to navigate down to the seas of adventure.

ARMED LINERS

To set on record the formation of the ships of the new navy in divisions, squadrons or units, and to classify them here under separate headings—an easy enough matter with regular fleets constructed for definite duties—is a task of considerable difficulty with a heterogeneous fleet composed of several thousand vessels with seldom two alike.

Beginning with the ocean liners, as the largest and most powerfully armed of the new fleet: these were mostly grouped for administrative purposes in one large formation, known as the "Tenth Cruiser Squadron." But when at sea they operated in smaller units and frequently as single ship patrols. Their principal zone of activity was the vast stretch of Arctic sea extending from Norway and North Russia to Iceland, the Hebrides and Labrador. Their work was arduous in the extreme, as will easily be realised from the nature of the seas in which they primarily operated.

Strictly speaking, were distinct divisions possible, the Tenth Cruiser Squadron did not form part of the auxiliary navy in its true sense, although many of the officers and men were drawn from newly raised corps. It acted rather as a distinct patrol fleet, filling the wide gap of sea between Scotland and the Arctic ice.

FIGHTING SLOOPS

Next in order of importance came the newly built screw sloops, with powerful guns and engines. Their numbers varied and they were continually being added to. Some of these vessels were used for patrol duties and others for minesweeping. The sloop flotillas had many zones of activity. One was the North Atlantic, with special care for the coast of Ireland. Another was the North Sea, with a marked preference for the east coast of Scotland and the Straits of Dover.

These flotillas also were frequently assigned duties independent of the auxiliary patrol organisation, but nevertheless formed an important part of the vast anti-submarine and anti-mine navy.

In the Mediterranean also there were a number of patrol gunboats and minesweepers similar to the fighting sloops. Their principal base in this region was on Italian soil.

Armed Yachts

We now come to that portion of the auxiliary fleet whose special care was the seas around the United Kingdom and the Colonies. First came the armed yachts, over 50 in number, with tonnages varying from one to five hundred. These were obtained from the owners, armed as heavily as their size and strength permitted, and mostly became the flag-ships of patrol flotillas. They were nearly always equipped with wireless, hydrophone listening apparatus, depth charges and all the appliances for anti-submarine warfare.

Their losses were not heavy considering the dangerous nature of their work and could almost be counted on the fingers of both hands. This was due mainly to their good speed and manoeuvring qualities. They made wonderfully efficient auxiliary warships, maintaining the sea in almost all weathers and accounting for quite a number of U-boats. These vessels were, of course, never used for the rougher work of minesweeping.

Whalers

The whalers were few in number and resembled small destroyers. They were powerful craft and well armed, but their sea-keeping qualities left much to be desired. In fact, to use a naval term, they were dirty boats even in a "lop." It was said that if an officer or man had been for long in one of these ships he was proof against all forms of sea-sickness. A big assertion, as even old sailors will admit—but they call it "liver."

MINESWEEPERS

About the screw and paddle minesweepers little can be said beyond the fact that they numbered about 200 and performed some of the most dangerous work in the war. Many of them were old passenger steamers from the Clyde, Bristol Channel, Thames and south and east coast resorts, the famous *Brighton Queen* being, until her untimely end on a mine off the Belgian coast, one of their number. The loss among this class of ship was about 10 per cent.

TRAWLERS

By far the largest portion of the auxiliary patrol units consisted of armed and commissioned trawlers. Their numbers far exceeded 1000, and nearly half were used for the dangerous work of minesweeping. About a trawler little need be said, for beyond what can be seen in the accompanying illustrations there is little of interest until we come to the question of their curious arms and appliances, fit subjects for a special chapter.

A large number of these units were fitted with wireless and carried masked batteries of quick-firing guns. To give here their zones of operation would be to set out in detail not only the seas around the British Isles, but distant waters such as the Mediterranean and the White Sea. They had distinct duties to perform, which may be summed up as follows:—(1) minesweeping; (2) night and day patrols alone or in company over immense areas of sea; (3) convoy duty; and (4) fishery guard.

Their losses were heavy, both in ships and men, amounting to about 30 per cent. Many were the lonely sea fights engaged in by these vessels. A few will receive the praise they deserve and the remainder will rest content with the knowledge of duty done.

If numbers or losses were the dominant factors the armed drifters should be high in the list. There were engaged considerably over 1000 of these craft, and the losses amounted to about 20 per cent.

It may be necessary to inform some of my readers that a drifter is not necessarily a vessel that is content to start out on a voyage and rely on *drifting* to its destination, as its name implies. The term is derived from the drift nets used by these vessels for fishing in time of peace. They are, in almost all respects, small editions of the deep-sea trawler—*minus* the powerful steam-driven winch for hauling in the trawl nets.

For war purposes the holds of these, and many other types of auxiliary warships, were converted into officers' cabins, or gun platforms for masked batteries. A few carried special nets in which to entangle the wily "Fritz." Others had aboard special types of submarine mines, and one, commanded by the author, was used for the transport of wounded from Admiral Sir David Beatty's flag-ship, H.M.S. *Lion*, after the Jutland fight.

These were, as might be expected, good sea boats, and carried out duties of great danger and value. Several hundred were fitted with wireless. Their zone of operations was far flung, extending from the Arctic Circle to the Equator. It was, however, in the unequal fights with German destroyers in the Straits of Dover and with Austrian torpedo boat destroyers in the Adriatic that they made a name for valour. In two of these engagements no less than six and fourteen drifters were sunk in a few minutes.

MOTOR LAUNCHES

About the now famous motor launches, or "movies," as they are called in the Service, much will be said in later pages. They numbered over 500, and, with but few exceptions, were a homogeneous flotilla of fast sea-going patrol boats, heavily armed for their size. Some idea of their appearance under varying conditions will be gained from a study of the illustrations.

They were all commanded by R.N.V.R. officers, whose training

SOME OF THE 550 MOTOR LAUNCH HULLS BEING CONSTRUCTED ON THE BANKS OF THE ST. LAWRENCE RIVER, CANADA

on H.M.S. *Hermione* and elsewhere has been described in an earlier chapter. They carried a crew of nine men and two officers, and their zones of operations extended from the icy seas which wash the Orkneys and Shetlands to the West Indies and the Suez Canal.

It may be of interest to give here an extract from the American journal, *Rudder*, showing how these vessels came into being.[3] Although the hulls were constructed in Canada, and much of the assembling was also carried out on the banks of the St Lawrence, the engines came from the United States. It was to the organising ability of Mr Henry R. Sutphen, of the Electric Boat Company, New York, that the delivery of over 500 of these wonderful little craft in less than a year was due. Here is that gentleman's story of the "M.L." contract:

It was in February, 1915, that we had our initial negotiations with the British Naval authorities. A well-known English shipbuilder and ordnance expert was in this country, presumably on secret business for the Admiralty, and I met him one afternoon at his hotel. Naturally the menace of the German submarine warfare came into discussion; we both agreed that the danger was a real one, and that steps should be taken to meet it.

I suggested the use of a number of small, speedy gasoline boats for use in attacking and destroying submarines. My idea was to have a mosquito fleet big enough to thoroughly patrol the coastal waters of Great Britain, each of them carrying a 13-lb. rapid-fire gun.

I explained that I had in mind two distinct types. The first would have an over-all length of about 50 feet, and would be fitted with high-speed engines; such a boat would show a maximum of 25 knots. The alternative would be something around 80 feet in length, with slow turning engines and a speed of 19 knots. I added that my preference was for the larger and slower type.

He asked how many units of that class we could build in a year's time, and I told him that I could guarantee fifty. He said that he would think the matter over, and we parted.

---

3. *Yachting Monthly* and *R.N.V.R. Magazine*, August, 1917.

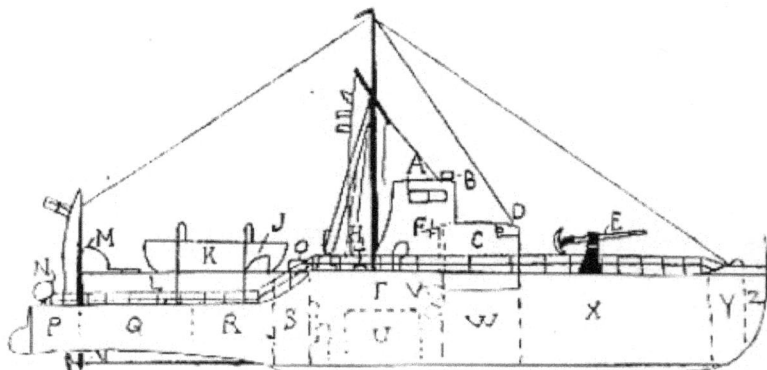

FIG. 1.

Diagram showing principal characteristics of an armed motor launch. *A*. Wheel-house. *B*. Searchlight. *C*. Chart-room. *D*. Navigation lights. *E*. 3 or 13 pounder quick-firing gun. *F*. Wheel and indicators in wheel-house. *H*. Hand pumps supplementing power pumps in engine-room. *I*. Hatchway leading to engine-room. *J*. Hatchway leading to wardroom. *K*. Life-boat. *L*. Officers' cabins. *M*. Hatchway leading to officers' cabins. *N*. Depth charges (2 or 4). *O*. Deck box containing life-belts. *P*. Stern petrol tanks (2). *Q*. Officers' sleeping cabin. *R*. Officers' mess-room. *S*. Galley. *T*. Engine-room. *U*. Main petrol engines (2). *V*. Reservoirs of compressed air for starting main engines. *W*. Forward petrol tanks. *X*. Forecastle and men's quarters. *Y*. Men's lavatory. *Z*. Forepeak.

A few days later I had another interview and was told that the British Government was ready to give us a contract for fifty vessels of the larger type, the whole lot to be delivered within a year's time. On April 9th, 1915, the contract for fifty 'chasers' was signed.

****

The *Lusitania* sailed on her last voyage May 1st, 1915, and a week later her torpedoing by a German U-boat was reported. My English friend was sailing that same day from New York, and we were giving him a farewell luncheon at Delmonico's. When the appalling news was communicated to him he appeared much depressed, as indeed was natural enough, and also very thoughtful. Before he said good-bye he intimated to me that he intended advising the Admiralty to increase the number of 'Chasers'; he asked me if I thought I could take care of a bigger order. I told him that I could guarantee to build a boat a day for so long a period as the Admiralty might care to name.

44

After he reached England we shortly received a cablegram ordering five hundred additional '*Sutphens*,' our code word for submarine 'Chaser'; in other words we were now asked to build five hundred and fifty of these boats and deliver them in complete running order by November 15th, 1915.

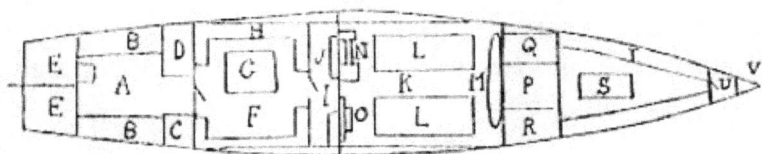

FIG. 2.

Plan of armed motor launch, showing internal arrangements. *A.* Officers' sleeping cabin. *B.B.* Bunks. *C.* Cupboard. *D.* Lavatory. E.E. Stern petrol tanks. *F.* Wardroom. *G.* Table. *H.* Settee. *I.* Galley. *J.* Petrol stove. *K.* Engine-room. *L.L.* Main engines. *M.* Compressed air reservoirs. *N.* Auxiliary petrol engine driving dynamo, bilge pumps, fire pumps and air compressor. *O.* Electric storage batteries, switchboard and electrical starting arrangements for auxiliary engine. *P.* Chart-room with petrol tanks below. *Q.* Magazine. *R.* Fresh-water tanks. *S.* Forecastle. *T.* Bunks for crew. *U.* Forecastle lavatory. *V.* Watertight forepeak.

The armament of a motor launch consisted of a 13-pounder quick-firing high-angle gun, capable of throwing a Lyddite shell for over four miles, and was as useful against aircraft as it was against submarines. In addition to this heavy gun for small craft they carried about 1200 lb. of high explosive in the form of depth charges for bombing under-water craft, a Lewis machine gun, rifles and revolvers.

These vessels were driven by twin screws connected to twin engines of about 500 h.p. They possessed, in addition, an auxiliary petrol engine of about 60 h.p. for compressing the air required to start the main engines, for working the fire and bilge pumps, and for driving a dynamo to recharge the electric storage batteries. The triple tanks carried over 3000 gallons of petrol, and the consumption, when travelling at full speed, was a gallon a minute.

Many were fitted with wireless, and all of them had on board the most approved pattern of hydrophone, with which to listen below the surface for the movements of hostile submarines. They had electric light in the cabins and for navigation, fighting and mast-head signalling purposes. A moderately powerful searchlight, fitted with a Morse signalling shutter, was also part of their equipment.

These little miniature warships possessed a small wardroom and sleeping cabin for the officers, a galley with petrol range for cooking, an engine-room, magazine for the ammunition, chart-room, and ample forecastle accommodation for the crew of nine men. All parts of the ship were connected with the bridge by speaking-tubes and electric bells, and the aft deck accommodated a steel life-boat.

The duties of these craft varied considerably. For over three years they maintained a constant patrol in the North Sea, Atlantic, English Channel, Irish Sea, Mediterranean, Adriatic, Suez Canal, Straits of Gibraltar, and in West Indian waters. Only one who knows by experience can fully appreciate what work in these northern seas, with their winter snows and Arctic winds, and their chilly summer fogs, really means to a mere thirty tons of nautical humanity in as many square leagues of storm-swept sea infested with mines and hostile submarines. But when this book has been finished the reader will be in a position to judge for himself.

The losses of motor launches were not heavy considering the dangerous nature of their cargoes (3000 gallons of petrol within a few feet of 1500 lb. of high explosive in a wooden hull) and the duties they were called upon to perform in all weathers short of heavy gales. Several were blown up with terrible results to those aboard. Others caught fire and were burned—allowing only just sufficient time to sink the explosives aboard. A few were smashed to pieces on exposed coasts after struggling for hours amid heavy seas. One struck a mine off Ostend. Another was destroyed by shell-fire in the Mediterranean, and the part they played in the raids on Zeebrugge and Ostend, in which two were lost and a V.C. gained, is now world famous.

Coastal Motor Boats

There was, besides M.L.'s, another smaller but faster type of submarine chaser. These little vessels, of which there were about 80 actually in commission, possessed no cabin or other accommodation for long cruises. They were simply thin grey hulls with powerful high-speed engines. They were known as C.M.B.'s, or, to give them their full title, Coastal Motor Boats. The purpose for which they were constructed was to operate from coastal bases, and to be launched from ocean-going ships to chase a hostile submarine which had been located by seaplanes and reported by wireless in a given locality. This, however, was what they were *intended* for, but bore little relation to the work they actually accomplished. Their nickname was "Scooters," and they certainly did "scoot" over the sea.

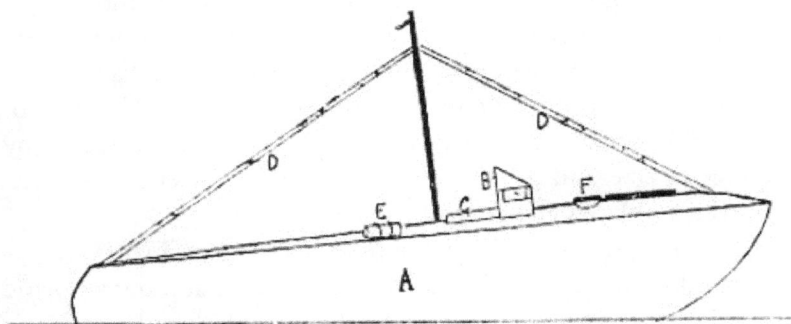

Fig. 3.

Diagram showing principal characteristics of a coastal motor boat (C.M.B.). Speed 50 miles per hour. *A.* Hydroplane hull, so constructed as to rise on to surface when travelling at full speed. *B.* Covered wheel-house. *C.* Navigating well. *D.* Wireless aerials. *E.* Depth charges (2 small size). *F.* Manhole to engine-room.

There were three types of C.M.B.'s. One had a length of only 44 feet, and was intended for carriage on the decks of light cruisers or other moderate-sized surface ships. The armament was a Lewis machine gun and two depth charges for anti-submarine warfare. The next class were 55 feet in length and operated from coast bases. These were fitted with one or more Whitehead torpedoes, launched by an ingenious contrivance from the stern. Class III. were 70 feet in length, and were commissioned just before the signing of the Armistice. They were fitted for mine-laying close up to enemy harbours.

The maximum speed of the 55-feet C.M.B.'s, which were the most numerous, was 40 knots, or nearly a mile a minute. They were driven by twin screws coupled to twin engines of 350 h.p.

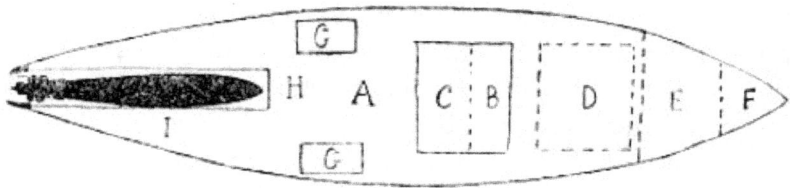

FIG. 4.

Plan of coastal motor boat, showing torpedo in cleft stern. *A*. Whale-back or arched deck. *B*. Wheel-house. *C*. Navigating well. *D*. Engine-room. *E*. Forward petrol tanks. *F.* Forepeak. *G*. Depth charges. *H*. Cleft stern with torpedo ready for launching. *I*. Whitehead torpedo, launched stern first.

each—working at 1350 revolutions per minute. Being of very shallow draught, some 26 inches, these little vessels could skim, hydroplane fashion, over any ordinary mine-field, and a torpedo fired at them would merely pass under their keel. The risk of destruction from shell-fire was also reduced to a minimum by their small size and great speed. Their principal enemies were, however, seaplanes armed with machine guns.

It is not difficult to imagine a fight between a C.M.B. travelling at 40 knots, firing with its little Lewis gun at a big seaplane swooping down from the clouds at the rate of 70 miles an hour, and splashing the water around the frail little grey-hulled scooter with bullets from its machine gun. This actually occurred many times off the Belgian coast, and is a typical picture of guerrilla war at sea in the twentieth century.

The C.M.B. was a purely British design, and the firm largely responsible for the success achieved was Messrs John J. Thornycroft & Company Limited. There were bases for these sea-gnats at Portsmouth, Dover, Dunkirk, and in the Thames Estuary at Osea Island. From all of these points mid-Channel could be reached in less than thirty minutes. Although useless in rough weather, a trip in a C.M.B., even on a calm day, was sufficiently exciting. The roar of the engines made speech impossible, and vision when sitting in the little glass-screened well, or conning-tower, was limited by the great waves of greenish-white water which curved upwards from either bow, and rolled astern in a welter of foam. There was an awe-inspiring fury in the thunder of the 700 h.p. engines revolving at 1350 per minute, and a

A 40-ft. COASTAL MOTOR BOAT TRAVELLING AT FULL SPEED

A 40-ft. COASTAL MOTOR BOAT TRAVELLING AT FULL SPEED

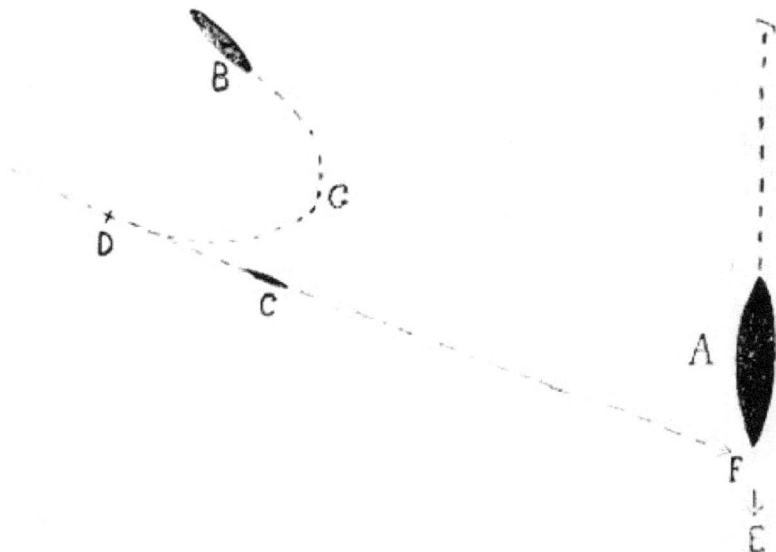

FIG. 5.

Diagram illustrating method of attack by C.M.B. on surface ship (or submarine on surface). *A.* Object of attack travelling in direction indicated by arrow *E. B.* The position of the C.M.B. after delivering the attack. *C.* The torpedo, released by the C.M.B. at point *D,* travelling on course ending at *F,* which, allowing for movement of ship *A,* is the place where the torpedo should strike its object of attack. From this it will be seen that the torpedo, when released, actually follows the ship from which it is fired until the latter swerves from the straight course, when the torpedo continues until it strikes or misses the object of attack, the speed of the torpedo being about the same or a little less than that of the C.M.B. The total time occupied in such an attack over a course of two miles would be about 2½ minutes before the torpedo struck its object.

feeling of ecstasy in the stiff breeze of passage and the atomised spray. When waves came the slap–slap–slap of the water as the sharp bows cleft through the crest and the little vessel was for a brief moment poised dizzily on the bosom of the swell caused tremors to pass through the thin grey hull, and, to complete the review of sensation, there may be added the human thrill of battle and the indescribable feeling of controlled power beneath one's feet.

The C.M.B.'s record of service, although short, is nevertheless a brilliant one. Towards the close of the year 1916 four of these little vessels coming from the base at Dunkirk intercepted five German destroyers returning from a Channel raid. The scooters raced towards the enemy in a smother of foam. Every quick-firing gun on the German ships spouted shells at the mysterious white streaks approaching them with the speed of lightning. So close did these plucky little ships go to their giant adversaries that the blast of the German guns was felt aboard, but no shells struck them. Then the line of C.M.B.'s swerved and their torpedoes were launched at close range. One of the enemy destroyers was hit and badly damaged, while two others had narrow shaves.

There was no time for German retaliation. For a brief few minutes the sea around the scooters was ploughed up by the shells from the Hun artillery, then the four little attacking craft were five miles distant from the scene of their victory, and presented almost invisible white specks to the enemy gunners.

At Zeebrugge these craft ran close in under the guns of the shore fortifications, and covered the approach of the landing parties and block-ships with a screen of artificial smoke. At Ostend they entered the harbour under heavy fire and ignited flares to enable the block-ships to navigate in the darkness. Others, in the same operations, torpedoed the piers and silenced the guns mounted thereon.

Their exploits savour of old-time sea romance, as, for example, when the little *Condor* ran in under the guns of the fortress of Alexandria, or further back in our naval history, when sail and round shot took the place of petrol and torpedoes.

For anti-submarine work these wonderfully fast little chasers were used in small flotillas. They were fitted with short-range wireless sets, and when the message came stating that a vessel was being attacked in a certain position, perhaps twenty miles from the coast, a number were instantly released from the leash, and in a fraction of the time taken by larger vessels they were on the scene with torpedoes and Lewis guns for surface attack and depth charges for submerged bombing.

They were commanded, in many instances, by R.N.V.R. officers of the auxiliary service, and carried two engineers. No crew was necessary, nor was space available for them. The plucky dash of these vessels into the harbours of Zeebrugge and Ostend, their subsequent operations on the Belgian coast, and their losses in the action at the entrance to the Heligoland Bight in 1918, when they were launched from a big ship, have earned for them high renown in naval history.

BOOM DEFENCE SHIPS

In addition to all these types of anti-submarine craft there were, forming part of the auxiliary fleet, over 300 ships, mostly trawlers and drifters, engaged in maintaining the great lines of boom defences, closing vast stretches of sheltered waters frequented by the battle fleets, and a considerable number of examination ships, staffed by interpreter officers, whose duty it was to examine all neutral shipping passing through the 10,000 miles of the blockade.

****

These, then, were the ships of the new navy, and their formation into flotillas, or units, was usually accomplished by grouping four or five vessels of similar type together under the command of the senior officer afloat—mostly a lieutenant R.N.R. or R.N.V.R. In the case of minesweepers the unit nearly always consisted of an even number of ships, because their work was carried out in pairs, and with M.L.'s it usually consisted of five boats, as this was the number required for the intricate tactical work of submarine chasing.

There were, of course, units from the United States, French, Japanese, Italian and Brazilian navies, in addition to the formidable British armada.

The auxiliary units were all based on one or other of the fifty odd war stations which encompassed not only the coasts of Great Britain and Ireland, but also the littoral of every land in our world-wide Empire. The numbers given here do not include the local fleets of purely colonial naval bases, nor the large flotillas of destroyers and "P" boats operating in home and foreign waters in conjunction with the auxiliary navy. If these were incorporated the anti-submarine fleets would be almost doubled.

Now that the reader is familiar with the *raison d'être* of the new navy, the personnel, the ships and their formation into fleets, the scope and limitations of their activity, and of the losses they sustained, the way is clear for a description of the curious weapons used, the mysteries of anti-submarine warfare, and the bases themselves before entering the zone of war and seeing something of the actual work of the auxiliary navy.

CHAPTER 5

The Hydrophone and
the Depth Charge

Of all the weapons used in the anti-submarine war the two most important were the hydrophone and the depth charge. They were employed in conjunction with each other and comprised the surface warship's principal means of offence against submarines operating beneath the surface.

The hydrophone resembles a delicate telephone. It is so constructed that when the instrument is lowered over the side of a ship into the sea any noise, such as the movement of a submarine's propellers, can be heard on deck by an operator listening at an ordinary telephone receiver connected to the submerged microphone by an electrified wire.

There were many different types of hydrophone in use during the Great War. So important was this instrument for the work of submarine hunting that money was spent in millions, and a corps of naval and civil experts were engaged for several years, bringing it to a state of efficiency. Each type introduced into the Service was an improvement on its predecessor, and there were different patterns for the use of almost each class of vessel. The fast destroyer required a different instrument to the slow-moving trawler. The motor launch could only employ successfully a totally different type to the submarine, and, to add to the difficulties, the German submarines themselves were generously supplied with similar instruments. The games of "hide-and-seek" played on and under the seas with the aid of this wonderful little instrument would have been distinctly amusing had men's lives—and often those of women and children—not been dependent upon the issue.

The portable hydrophone, used by some of the smaller and slower vessels of the auxiliary fleet, consisted of a microphone, or delicate mechanical ear, carefully guarded by metal discs from accidental damage, and connected to ear-pieces or ordinary telephone receivers by an electric wire which passed through a battery. Where the wire came in contact with the sea water it was heavily insulated and lightly armoured.

When it was required to use this instrument the vessel was stopped and the microphone lowered overboard to a depth of about 20 feet. This was the distance down from the surface at which submarine noises could be heard most distinctly. The operator on deck or in the cabin then adjusted the ear-pieces and

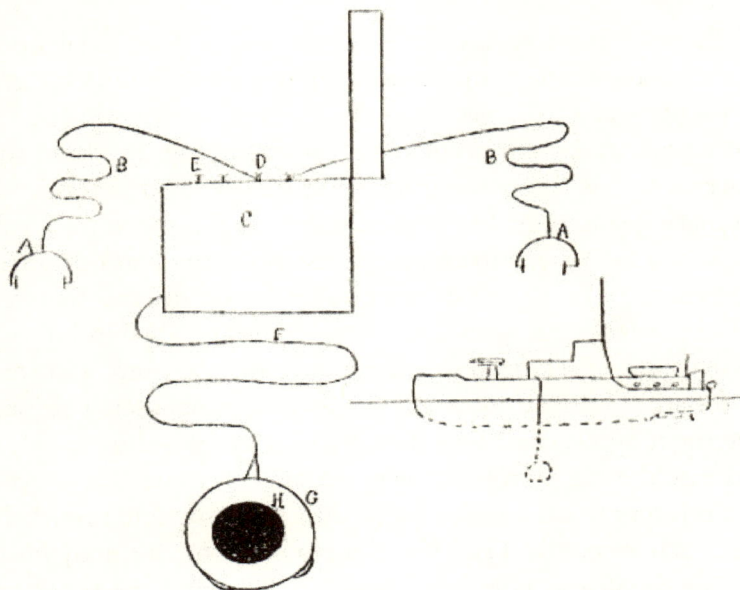

FIG. 6.

Diagram showing essential parts of a portable hydrophone. *A.* Head and ear pieces, by means of which a trained listener hears submarine sounds. *B.* Flexible leads to enable an officer to verify reports from listener. *C.* Battery box, containing spare set of cells. *D.* Terminals. *E.* Terminals of spare cells. *F.* Flexible armoured electric cable which is lowered over side of ship. *G.* Metal case protecting the microphone *H. H.* Microphone or delicate receiver of submarine sounds, which is submerged (when required, but not when ship is moving) to a depth of about 18 feet, as in small diagram. The sound is detected by the microphone and transmitted up the cable *F* and wires *B* to the ear-pieces *A*.

sat listening for any noises coming through the water. Although the sea is a far better conductor than air, the range at which sounds could be heard varied considerably. On a calm day or night the noise of a ship's propellers could frequently be distinguished at from five to seven miles; whereas on a rough day, with the sea splashing and the wind roaring, it was often difficult to hear anything beyond half-a-mile.

In fine weather a submarine could usually be heard at a distance of about two or three miles. There were, however, many microscopic noises of the under-seas which were picked up and magnified by this type of hydrophone. They were called "water noises," and often made it extremely difficult to differentiate between them and the sound of a moving submarine at a great distance. Later types were not so prone to these disturbing influences.

To describe here the different natural and artificial noises heard on a portable hydrophone is extremely difficult. One general statement can, however, be made. It is the noise caused by the rapidly revolving propellers of both surface ships and submarines that is the guiding factor in the work of detection by submarine sound. A destroyer travelling at full speed on a calm sea, when heard on a hydrophone resembles the roar of a gigantic dynamo. The sound does not alter as the distance between the *stationary* listening ship and the *fast-moving* warship increases or decreases; it continues to be a roar or low hum, according to distance, until it fades out of hearing altogether. The same statement applies also to a slow-moving cargo steamer, only in this case the *single* propeller is revolving very much slower, and, when listening on a hydrophone about two or three miles distant, each successive beat of the engines can be distinctly heard.

The simple movement of a vessel's hull through the water cannot be heard on a hydrophone. Therefore for detecting the presence in the vicinity of a *sailing* ship at night or in a thick fog this instrument is quite useless. The same drawback applies also to the location of a floating derelict or iceberg, and restricts the use of the hydrophone to faithfully reporting the presence of power-driven ships or special sound signals at a range of a few miles.

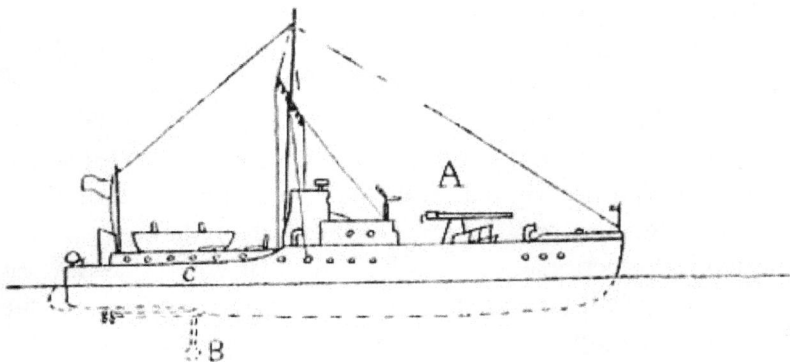

FIG. 7.

An improved directional hydrophone fitted through keel of motor launch. The tube *B*,
at the lower extremity of which is the microphone, can be raised or lowered from *C*,
the cabin of the M.L. This instrument is so arranged that the direction from which the
submarine sound is coming can be simply and quickly ascertained.

A German submarine heard at a range of about a mile on
a calm night presents a curious sound which almost defies description. Its principal constituent consists of a *"clankety clank!
clankety clank!"* at first barely distinguishable from the low swish
of the water past the face of the submerged microphone, then
louder when the sound has been distinguished and the human
ear is on the alert. But when this sound was heard in war there
was little time for analysing or noting. It was the call to action.
The microphone was hauled to the surface and the chase began,
a halt being made every half-mile or so for a further period of
listening on the hydrophone. If the sound was louder the commander of the pursuing vessel knew that he was on the right
track, and if the sound came up from the sea more indistinct the
course was changed and a run of a mile made in the opposite
direction, when the vessel was again stopped and the instrument
dropped overboard.

Should this manoeuvre have placed the surface ship in close
proximity to the submarine, one or more depth charges were released, and if the explosion of these damaged the comparatively
delicate hull or machinery of the under-water craft, she had
either to rise to the surface and fight for her life with her two
powerful deck guns, or, if badly damaged, sink helplessly to the
bottom, emitting oil in large quantities from her crushed tanks.

Before entering upon a description of the depth charge, however, there is more to say of the hydrophone, which has played such an important part in the defeat of the U-boats.

When the advantages of this instrument had been fully demonstrated in the stern trial of war, successful efforts were made to improve upon the original crude appliances. The "water noises" were reduced and, greatest improvement of all, the hydrophone was made "directional." By this is meant that when a sound was heard its approximate direction north, south, east or west of the listening ship could be more or less accurately determined. What this improvement meant to a vessel hunting a submarine in a vast stretch of sea will be easily realised. When the sound came up the wires from the submerged microphone the operator had simply to turn a small handle in order to determine from which direction the noise was coming.

If, for example, the sound was first heard away to the east, the instrument was turned to another quarter of the compass. Then, if the noise was plainer, the instrument was turned again until the sound decreased in intensity. In this way the line of maximum sound was obtained, and this showed the direction from the listening ship in which the U-boat was operating.

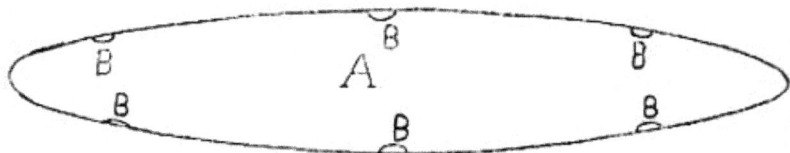

FIG. 8.

Plan showing how microphones or ears B are fitted in a submarine A to enable it to detect the approach of surface craft.

With the perfection of this invention the hydrophone section of the naval service came into being. Special courses in the detection of submarine sounds were instituted for officers and also for seamen listeners. The actual movements of a submarine under water at varying distances from a hydrophone were recorded by a phonograph, and records made so that the sounds might be reproduced at will for the education of the ear. Surgeons with aural experience estimated the physical efficiency in this respect of would-be volunteers for the hydrophone-listening service, and vessels were formed into special hydrophone flotillas, whose duties consisted of listening in long lines for submarines and when a discovery was made attacking them in the most approved tactical formation, with the aid of depth charges and guns.

A considerable measure of success attended these arrangements, and the author spent many cold hours listening at night for the sound of the wily submarine. On more than one occasion an exciting chase resulted.

It must, however, be pointed out that there is one great drawback to the successful use of the hydrophone. It exists in the necessity for the listening ship to stop before the hydrophone is hoisted outboard, it being quite impossible to hear anything beyond the roar of the engines of the carrying ship so long as they are in motion. Furthermore, all progress through the water must have ceased and the listening ship have become stationary before artificial sounds, such as the propellers of a submarine, can be distinguished from the natural noises of the sea water.

Now it will at once be apparent that not only does a stationary ship offer a splendid target for under-water attack, but also it allows a somewhat humorous game of hide-and-seek to be played between a hunting vessel and a hunted submarine.

Nearly all U-boats were fitted with a number of hydrophones and therefore were as well able to receive timely warning of an approaching surface ship as the surface ship was of the presence of the submarine. But the surface ship had the advantage of speed.

The result of all this was that when a German submarine heard a surface vessel approaching she dived to the bottom, if the water was not too deep or the sea-bed too rocky. Then shutting off her engines she listened. The surface ship, mystified by the sudden cessation of the noise she had been pursuing, also waited, and this stagnation sometimes lasted for hours. Then if the surface ship moved, as she was often compelled to do in order to avoid drifting with the tide away from the locality, the submarine moved also, and the one that stopped her engines first detected the other, but could not catch up to her again without deafening her own listening appliance. In which case the next move would probably be in favour of her opponent.

All of this is, perhaps, a little complicated, but a moment's pause for reflection will make this curious situation clear to the reader. And so the game went on, with decisive advantage to neither the surface ship nor the submarine. Darkness usually intervened and put an end to further manoeuvring, frequently allowing the submarine to escape.

A case of this kind occurred to a vessel, of a certain hydrophone flotilla, commanded by the author. For over four hours the U-boat eluded the pursuing surface ships by moving only when they moved and stopping when they too had stopped, darkness and a rising sea eventually favouring the escape of the submarine, which, a few hours later, was able to attack (unsuccessfully) a big surface ship less than thirty miles distant from the scene.

Nevertheless the hydrophone is a submarine instrument with a brilliant future. It has already been improved out of all resemblance to its original self, and more will undoubtedly follow. It is, however, purely an appliance for the detection of submarines when cruising beneath the surface, and not a weapon for their destruction. It should also be remembered that any improvement made in the efficiency of the hydrophone will benefit not only the surface ship, but also the submarine, for it cannot be supposed that under-water craft will be left without these wonderful submarine ears when their surface destroyers are equipped with them.

The alliance between the hydrophone and the depth charge is a natural one. The former instrument enables the surface ship to discover, first, the presence of a submarine in the vicinity, and, secondly, its approximate position. At this point its utility *temporarily* ceases and that of the depth charge begins. When a surface ship is hot on the track of a moving submarine she endeavours to attain a position directly over the top of her quarry, or even a little ahead, and then releases one or more depth charges according to whether the chance of a hit is good or only poor.

From this it will be apparent that whereas the hydrophone is the instrument used for the initial detection of the submarine, and afterwards for enabling the surface ship to get to close quarters with her submerged adversary, it is the depth charge with which the attack is actually made.

This weapon is really a powerful submarine bomb. It consists of several hundred pounds of very high explosive encased in a steel shell, with a special firing device which can quickly be set so that the charge explodes at almost any depth below the surface after being released from the above-water vessel.

The methods in use during the war for its release from the decks of surface ships were very diverse, the most usual being for a number of these weapons to be fitted on slides and held in place by wire slings which could be released by simply pulling out a greased pin or bolt.

When the depth charge rolled off the stern of the surface ship it sank to the "set depth" and then exploded like a submarine mine. The result was a shattering effect exerted through the water for several hundred feet around. If the submarine was close to the explosion her comparatively thin plates were nearly always stove-in. When she was over a hundred feet away, however, the rivets holding her plates together were often loosened,

DROPPING DEPTH CHARGES

and the resulting leak frequently compelled her to come to the surface, where she could be destroyed by gun-fire.

It often happened, however, that neither one nor the other of these things occurred, but that the submarine's delicate electrical machinery was thrown out of order by the violence of a depth-charge explosion, even when a considerable distance away. With the electric engines used for submerged propulsion no longer available, and possibly the interior of the vessel in darkness, there were only two courses open. She could either rise to the surface and endeavour to fight it out with the aid of her powerful deck guns, or else sink to the bottom and trust to luck that other depth charges would not be dropped close enough to seriously damage her hull. In the open sea, however, the latter chance was denied because of the depth of water. Three hundred feet

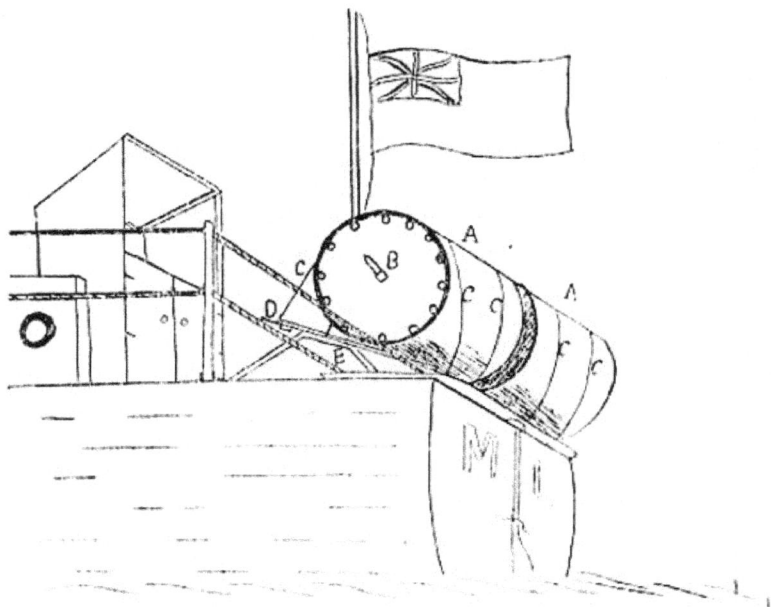

Fig. 9.

Diagram showing how depth charges are carried on the stern of a motor launch. *AA.* Depth charges, each containing 300 lb. of high explosive. *B.* Hydrostatic device by means of which the charge can be made to explode when it has sunk from the surface to a depth of 40 or 80 feet, and by which it is rendered comparatively safe while on deck. *C.* Slings holding charges in place on inclined platform. *D.* Greased bolts which, on being pulled out, allow wire slings to fly free and depth charge to roll into the sea. Depth charges can only be released from vessels under way, otherwise the explosion which occurs a few seconds after release damages surface vessel.

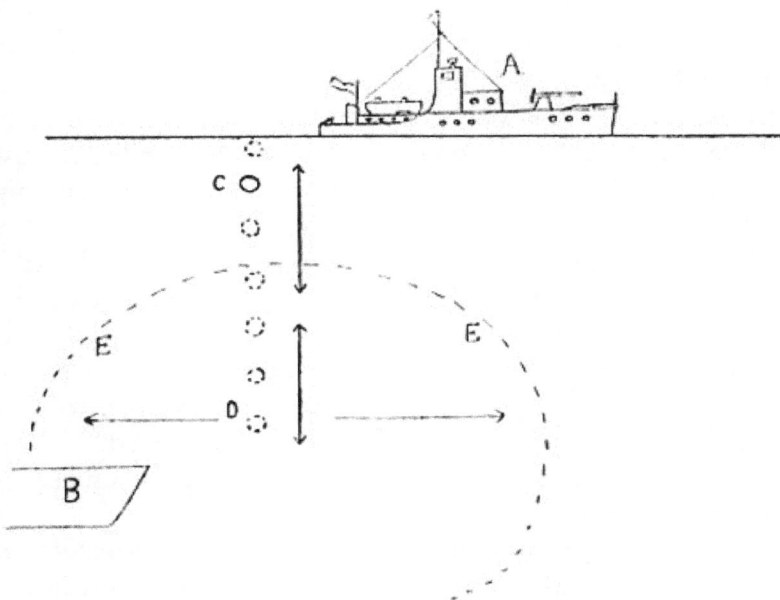

Fig. 10.

Diagram illustrating a depth charge attack on a submerged submarine. *A*. Motor launch, which has dropped a depth charge to destroy a submarine *B* travelling at a depth of 90 feet below the surface. *C* is the depth charge sinking as the M.L. steams away from the danger area. *D* is the point (80 feet below the surface) at which it will explode, and *E* indicates the danger area for the submarine *B*.

may be taken as the greatest depth to which an ordinarily constructed fighting submarine can safely descend without running a grave risk of having her plates crushed in by the great water pressure. Even at this depth the weight on every square foot of hull surface exceeds 8¾ tons.

If the damaged submarine rose to the surface the guns of her pursuers were ready and could generally be relied upon to place her at least *hors de combat* before the hatches of the under-water vessel could be opened and her own guns brought into action.

In shallow water where there was a fairly smooth bottom it generally happened that a submarine damaged by depth charges elected to sink to the sea-bed and trust to luck. This was also frequently resorted to as a means of eluding pursuit even when the U-boat was not damaged by the first few charges dropped. It was then that the hydrophones carried by the surface ships were again brought into use to ascertain if the submarine was still under way. When no sound was heard those on the surface knew that "Fritz" was lying doggo, or else that he had escaped. If a number of ships were available a few waited over the spot where it was considered the U-boat was lying, while the others scoured the surrounding seas in circles trying to pick up the sound of the runaway's engines if she had escaped in the mêlée. When nothing further was heard they returned to the scene and set about the work of systematically bombing the surrounding sea-bed.

As many as one hundred depth charges were dropped in quite a small area of sea and yet a submarine known to have been lying "doggo" in the locality was not damaged. In cases such as this other means, which will be described in a succeeding chapter, were then resorted to.

All the foregoing sounds very thorough and hopeful, but in fairness it must be said that submarine hunting is a heartbreaking task. The reader may have noticed that the method of depth-charge attack pre-supposes the surface vessel to have attained a position almost directly over the top of her enemy, a manoeuvre extremely difficult of achievement even with the most efficient hydrophone. Heavy seas, snow and fog have also to be taken into consideration, to say nothing of darkness, the presence of a second submarine, a surf-beaten rock or sandbank and the confusing sounds of passing merchant ships, making a difficult task more difficult, as will be seen when we come to the actual fighting.

CHAPTER 6

Some Curious Weapons of Anti-Submarine Warfare

Although modern war has shown that there exists no certain antidote for the submarine, it nevertheless brought into being many curious weapons of attack and defence. It is the purpose of this chapter to describe some of the anti-submarine devices used with more or less successful results during the protracted naval operations against the Central Powers.

INDICATOR NETS

Among the most important of these were the immense meshes of wire known as "indicator nets," which were used to entangle a submarine and then to proclaim her movements to surface ships waiting to attack with guns and depth charges.

These nets were made of specially light but strong wire, with a mesh of several feet. They were joined together in lengths of 100 feet by metal clips which opened when a certain strain was exerted on any particular section. Their depth was usually about 50 feet, and they were laid in lengths varying from a few hundred yards to two miles. Weights at the lower end and invisible glass floats along the top held them suspended vertically from the surface. The floats were kept in place by a wire hawser running along the top of the nets, and to this were attached, at intervals, wooden buoys containing tin cases filled with a chemical compound which, when brought into contact with sea-water, emitted dense smoke by day and flame by night.

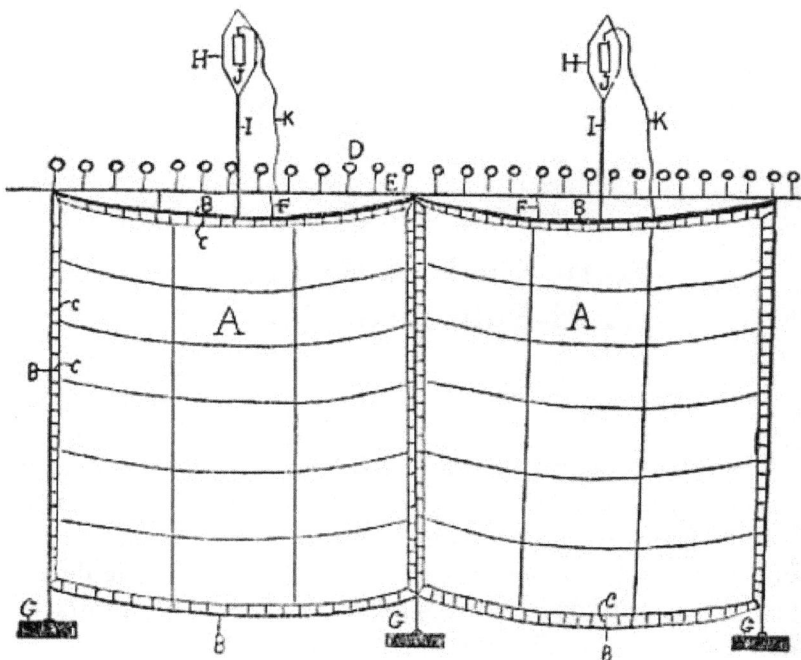

FIG. 11.

Diagram showing principal features of a line of submerged indicator nets. *AA*. Two sections (100 feet in breadth) of thin wire-netting with a very wide mesh. *B*. Framework of wire rope holding each section of net in place by means of metal clips *C*. *C*. Metal clips which expand and release netting from rope frame when a pull of more than 100 lb. is exerted upon them. *D*. Line of invisible glass balls, or hollow floats, attached to a surface wire *E*, supporting by wires *F*, the nets which hang down from the surface vertically in long lines (½ to 1 mile in length and 50 feet deep). *G*. Heavy iron weights or sinkers holding down the nets by their weight when hanging in water. *H*. Wooden floats, attached to each section of net by wires *I.J*. Canisters of chemical which give off flame and smoke when exposed to sea-water. *K*. Lanyard attached to surface wire *E*. When a section of net is pulled out of its wire frame by a submarine passing through the line the float is dragged along the surface by the wire *I*. The lanyard is held back by being attached to surface wire *E*, and pulls a plug out of the canister *J*, exposing the chemical inside to the sea-water (see Fig. 12).

The 100-feet sections were linked together, and to the top and bottom ropes, by the metal clips. These clips opened when a submarine headed into that part of the line. The result was that a section of net enveloped the bow of the under-water craft, was detached from the line and carried along, dragging its *indicator float* on the surface behind.

The indicator float, containing the chemical, was attached

(1) to the section of net by a short wire and (2) to the top rope of the whole line by a lanyard, which, when pulled free, exposed the chemical contents of the canisters in the float to the sea-water. The float was then dragged along the surface burning furiously.

As there was nothing to materially impede her progress, a submarine would consequently be unaware that she had passed through a line of nets and was actually towing a flaming buoy. Even if she became aware of the tell-tale appendage it would be extremely difficult to clear herself, owing to the forward hydroplanes becoming entangled in the wire-netting, before the fast surface ships, waiting in readiness, had spotted the flaming buoy being towed along and were hot in pursuit.

Once entangled in such a net, the submarine's chance of avoiding destruction was small. Not only did the indicator buoy proclaim her every movement to the pursuing surface ships, so that she could not avoid them by turning, sinking to the bottom or doubling in her tracks, but it also enabled depth charges to be literally dropped on her decks.

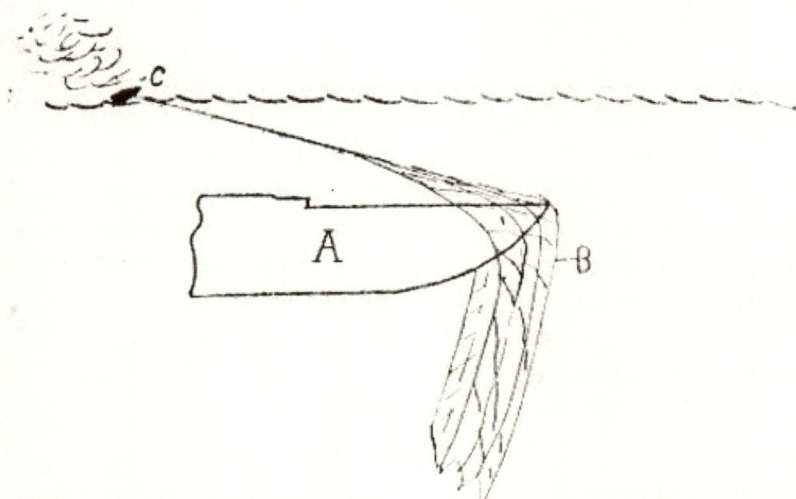

Fig. 12.

Diagram showing a submarine entangled in a submerged net. The submarine A after passing through a line of nets emerges with her bows enveloped by one section B which she has carried out of its wire-rope frame. The flaming buoy C, betraying her movements, is being towed along the surface.

A considerable measure of success attended the use of this ingenious device until "Fritz" became shy of waters close inshore, and kept a careful look-out for possible lines of indicator nets when forced to pass through narrow channels and waterways. One of the main disadvantages attending the use of these nets was the impossibility of laying them—or, when laid, of hauling them inboard again, during even moderately rough seas. Another difficulty which presented itself when indicator nets were required to be laid in the open sea was the screening of the waiting surface ships from observation. Submarines could not be used on account of their slow speed, and when fast patrol craft cruised about openly within easy range of the nets "Fritz" suspected a trap and steered clear. Even this, however, had its uses.

Mine Nets

It was sought to overcome this difficulty by attaching small explosive mines to the nets instead of indicator floats, so that when a submarine passed through a line she unavoidably struck one or other of the attached mines, which instantly exploded.

This device also proved fairly successful, but the dangers of handling mined nets were considerable and disasters resulted. Furthermore, as such obstructions could not be securely moored in one spot for very long, owing to the action of gales and strong tides, it became necessary for the sake of neutral and allied shipping to maintain a vessel in the vicinity from which warnings could be issued and repairs to the nets effected. This partly defeated the object of mined nets, except for the closing of narrow fair-ways, and their scope as a weapon of attack became strictly limited.

The Modified Sweep

This elaborate and costly anti-submarine device was very widely, but not altogether successfully, employed by the auxiliary fleet during the first two years of war. It was nothing more than a long explosive tail towed submerged by a surface ship,

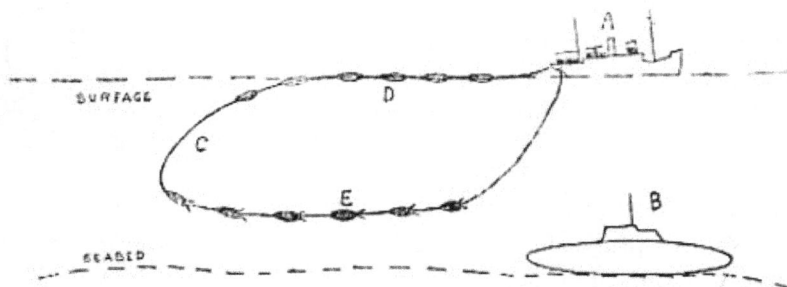

FIG. 13.

Diagram showing a vessel towing a modified sweep. This appliance consists of an armoured electric cable G towed in vertical loop under the surface. The floats D support the 100-lb. charges E, which have strikers attached. If a submarine B is lying "doggo" on the sea-bed one or other of these charges may strike her hull and the whole line then blows up, shattering everything in the surrounding sea. If the strikers fitted on the charges do not touch the submarine the whole line can be exploded at will from the surface ship by closing an electric circuit.

the object being to either drag it over a submarine resting on the sea-bed, or else, if the under-water craft was moving, to so manoeuvre the towing surface ship as to swing the tail close to the U-boat, when the heavy charges of T.N.T. attached to the armoured electric cable, forming the tail, would be exploded either by actual contact with the hull of the enemy, or, when sufficiently close to be effective, by the closing of a firing circuit on board the surface ship.

Excellent in theory but very difficult of accomplishment in actual practice. The diagram given will explain the details of this elaborate contrivance, which, however, was soon discarded for more practical methods, although at least one German submarine is known to have been destroyed by it.

LANCE BOMBS

These little engines of destruction were intended for fighting at close quarters, and can be described here in a few lines because of their guileless simplicity. They consisted of conical explosive bombs on the ends of broom handles! A strong man could whirl one of them round his head, like a two-handed sword or battle-axe, and, when the momentum was sufficient, hurl it over the water for about seventy-five feet. On nose-div-

FIG. 14.

A lance bomb. The wooden handle *A* enables the charge *B* (7 lb. of high explosive) to be whirled round the head and hurled a distance of about twenty yards.

ing into the sea and hitting the hull of a submarine in the act of rising or plunging, the little bomb, containing about 7 lb. of amatol, was exploded by contact.

The damage inflicted on one of the earlier types of submarines by an under-water hand-grenade or lance bomb depended entirely upon what part of the vessel happened to be struck. Their sphere of usefulness was, from the first, very limited, and the advent of the big cruiser submarine, with armoured conning-tower and 5-inch guns, rendered them obsolete.

SMOKE SCREENS

We now come to a more useful device of the purely defensive type employed to screen surface ships from submarine attack. The very simple mechanical and chemical apparatus needed for making the heavy clouds of smoke needs no description beyond that given in the text, but something must be said here regarding the methods of use.

It was not until the third year of the Great War had been ushered in by the unprecedented sinking of Allied merchantmen by German U-boats that the value of the smoke screen as a means of baffling an under-water attack was fully realised. Convoy guards were supplied with the necessary appliances for emitting the fumes with which to cover the movements of the ships under their protection, and so successful was this method of blinding attacking submarines that within a few months thousands of transports, food-ships and warships had been equipped.

When a submarine proclaimed her presence in the vicinity of a convoy either by showing too much of her periscope or by a misdirected torpedo, the guard-ships on the flank attacked immediately dropped their smoke buoys as they continued moving at full speed. By this means an impenetrable optical barrier was interposed between the attacking submarine and the fleet of merchantmen under convoy. When thus shielded from attack—a submarine values her small stock of torpedoes (six to ten) too highly to risk the loss of one or more on something she cannot even see—the mercantile fleet altered course so as to present their sterns to the attacking U-boat, while certain prearranged warships belonging to the escort proceeded to the attack with guns and depth charges. This means of masking the movements of ships—by no means new in naval warfare—was employed with conspicuous success in the operations of Allied squadrons off Zeebrugge. Individual merchantmen, when attacked by one or more submarines, often threw out a smoke screen to avoid destruction by the big surface guns of the more modern German craft, and its use to cover the movements of transports was very frequently resorted to.

CAMOUFLAGE

The use of camouflage, or the deceptive painting and rigging of ships, came first into being owing to the method employed by submarines for judging the speed of passing surface ships by the white wave thrown off from their bows. It is of the utmost importance for the commander of an under-water warship to correctly judge the speed of the vessel he is about to attack before discharging a torpedo at her. If the estimated speed is too high the torpedo will, in all probability, pass ahead of the moving target, and if it is too low it will run harmlessly astern.

To cause this to happen as frequently as possible, and valuable

FIG. 15.

A camouflaged ship. It will be observed that a vessel so painted would, from a distance of several miles, give the appearance of a ship sinking while headed in the opposite direction.

torpedoes to be wasted—even if the attacking submarine herself could not then be discovered and destroyed—it became advisable to paint imitation white waves on the bows of slow-moving ships in order to give the appearance of speed.

So successful was this simple form of deceptive paint-work that a special camouflage section of the naval service, with an eminent artist as its director, was formed, and all kinds of grotesque designs were painted on the broadsides and superstructures of almost every British merchantman operating in the submarine danger zone.

There was method and meaning in the seemingly haphazard streaks of black, green, blue and white. When looked at from close range only a jumble of colours could at first be seen, but if the distance was increased the effect became instantly apparent. In some cases the deceptive decoration caused big ocean liners to appear small and insignificant. In others it gave the appearance that the vessel was sinking; while quite a favourite ruse was to cause the vessel to appear as if she was travelling in the opposite direction to that which she really was. Two-funnelled ships became single-funnelled, when viewed from a distance or in a dim light, by the simple expedient of painting one funnel black and the other light grey. Liners with tiers of passenger decks had the latter obscured by contrasts of colouring which were really masterpieces of deceptive art. In fact so deceptive became almost every ship in the dim light of dawn and dusk that collisions were often narrowly averted.

It frequently occurred that paint alone was not sufficient to disguise a ship, and woodwork and canvas were resorted to. Big guns were made of drain-pipes and shields of the wood from packing-cases. Cargo boats were given the appearance of cruisers, and cruisers reduced to the appearance of cargo boats. In this way hostile submarines were induced to attack ships, thinking them unarmed and helpless, when in reality they were small floating forts. But at this point simple camouflage ceases and the famous *Mystery Ship* begins. Before closing this chapter, however, it must be pointed out that camouflage only came into being when the German U-boats commenced their ruthless submarine warfare.

CHAPTER 7

Mystery Ships

The "Q" boat, or mystery ship, has been surrounded by so much secrecy that to most people its very being is an unknown quantity. Yet it is to these curious vessels of all sizes and types that the destruction of many hostile submarines was due, and the dangerous work performed by their intrepid crews equalled anything described in sea romance.

The mystery ship was not a specially constructed war vessel, such as a destroyer or cruiser, but merely a merchantman converted into a powerfully armed patrol ship, camouflaged to give the appearance of genuine innocence, but with masked batteries, hulls stuffed with wood to render them almost unsinkable, hidden torpedo tubes, picked gunners, a roving commission and a daring commander and crew. Their work was performed on the broad highways of the sea, and they hunted singly or in pairs, often fighting against overwhelming odds with certain death as the price of failure.

As all "Q" boats—as they were officially called—differed from each other in size and armament, any description given here can only be taken as applying to one or more vessels with which the writer was personally familiar. Some of these so-called mystery ships were old sailing schooners, others fine steamships, while quite a number were converted fishing smacks, drifters and trawlers, the method being to give the prospective commander a free hand in the conversion of his ship from a peaceable merchantman to a camouflaged man-of-war, and many were the ingenious devices used.

INNOCENT LOOKING BUT DEADLY, H.M.S. *HYDERABAD*

The famous "Mystery Ship," powerfully built to resemble a helpless merchantman. Sitting almost flat on the surface of the sea the torpedoes from U-boats ran harmlessly beneath her keel.

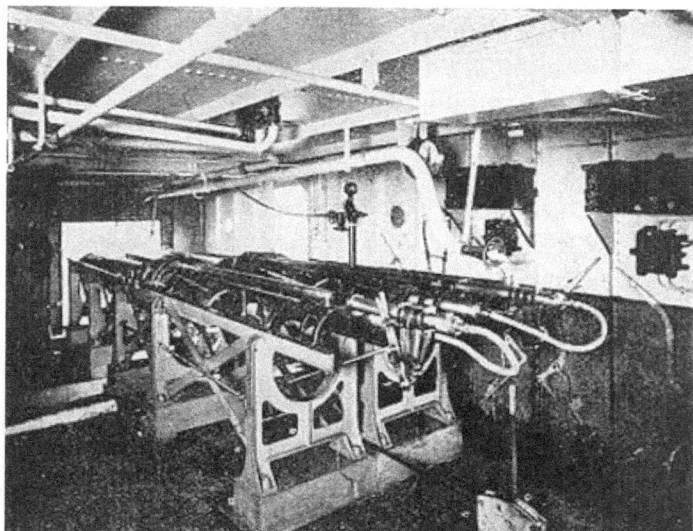

THE HIDDEN TORPEDO TUBES OF H.M.S. *HYDERABAD*

The number of these vessels was not large, possibly 180, but their operations extended far and wide. They roamed the North Sea, the Atlantic, the English Channel, the Mediterranean, the Arctic Ocean and even the Baltic, but until challenged were quite unknown to all other vessels of the Allied navies. Theirs was a secret service, performed amidst great hardships, with no popular applause to spur them on.

One vessel fitted out for this desperate duty at a Scottish base was a steamer of about 400 tons burden. She was armed with a 4.7 quick-firing gun hidden in a deck-house with imitation glass windows, the sides of which could be dropped flat on to the deck for the gun to be trained outboard by simply pressing an electric button on the steamer's bridge. Two life-boats, one on each side of the aft deck, were bottomless, and formed covers for two additional 12-pounder guns. A false deck in the bow shielded a pair of wicked-looking torpedo tubes, each containing an 18-inch Whitehead ready for launching; and the crew for each gun were able to reach their respective weapons, without appearing on deck, by means of specially constructed gangways and hatches. The very act of dropping the sides of the aft gun-house hoisted the White Ensign, and technically converted this unsuspicious-looking merchantman, which asked only to be allowed to pursue its lawful vocation on the high seas, into a heavily armed warship.

This "Q" boat had, when met and challenged by the writer's ship, already accounted for no less than three German submarines which had opened the attack from close range, thinking her defenceless.

Another smaller mystery ship was a converted fishing drifter with a single 12-pounder gun on a specially strengthened platform fitted in the fish-hold, which had been cleaned, match-boarded and painted to provide accommodation for the crew of picked gunners. This little ship had no torpedo tubes and the muzzle of her gun was hidden beneath fishing nets.

There were, however, some very large and elaborately fitted "Q" boats. These had specially constructed torpedo tubes low down in the hull, masked 4.7-inch guns in more than one position, special chutes for depth charges, coal bunkers arranged round the vital machinery to protect it from shell-fire, and, moreover, were filled with wood to make them almost unsinkable even if torpedoed.

Each such vessel was provided with a "panic party," whose duty was to rush to the life-boats when the ship was attacked by a submarine. This gave the final touch to the disguise, and often

75

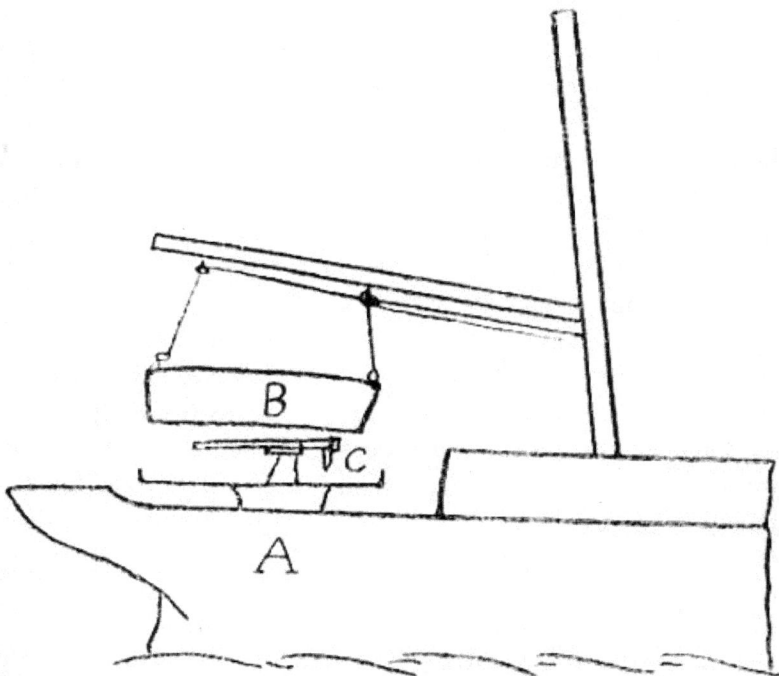

Fig. 16.
Method of masking a 3, 6, 12 or 13 pounder gun. *A*. Stern of ship. *B*. Shield constructed to resemble a life-boat which can be raised or lowered over gun *C*.

induced the submarine to save further torpedoes by coming to the surface and continuing the assault with gun-fire.

The story of the sinking of the last German submarine in the war by the "Q 19" will give some idea of how these vessels worked. It occurred in the Straits of Gibraltar, about twenty-four hours before the signing of the Armistice. The Q 19 was waiting in the Straits expecting to intercept three big U-boats on their way back to Heligoland. About midnight the first of these craft came along, and sighting the innocent-looking "Q" boat prepared to attack her with gun-fire. For nearly an hour the mystery ship "played" the submarine by pretending to make frantic efforts to escape, but all the time allowing the under-water craft to draw closer and closer.

The "Q" boat was under a heavy fire from the submarine, one shell wounding eleven out of the crew of sixty, another carrying away the mast and a portion of the funnel, but no sign of a gun was yet displayed on board the surface ship. This withholding of fire until the last moment, when the range has become short and the effect certain, is one of the great nerve tests imposed on the crews of all mystery ships. It is an essential of success, for a few wild shots at long range would disclose the fact that the vessel was heavily armed, and the attacking submarine would either sheer off or else submerge and use her torpedoes.

When the chase had been on for about fifty minutes, and the submarine was only 200 yards astern, the "panic party" in the "Q" boat rushed for the life-boats. The shells were now doing serious damage to both hull and upper works, and the submarine was creeping close to give the *coup de grâce*.

At this, the psychological moment, the order to open fire was given. The collapsible deck-house, shielding the 4.7 gun, fell away on its hinges. Eleven shots were fired in quick succession, all of which struck the submarine. One blew the commander off the conning-tower and another rent a gaping hole in the vessel's hull. In less than fifteen minutes the fight was over and the last U-boat to be sunk in the Great War of civilisation had disappeared beneath the waters of the Straits of Gibraltar.[4]

---

4. One of the remaining U-boats afterwards succeeded in torpedoing the battleship *Britannia*.

Chapter 8

A Typical War Base

The last few chapters have dealt mainly with the weapons used in anti-submarine warfare. We now come to the naval bases on which the fleets armed with these curious devices were stationed for active operations.

Around the coasts of the British Isles there were about forty of these war bases, each with its own patrol flotillas, minesweeping units and hunting squadrons. The harbours, breakwaters and docks had to be furnished with stores, workshops, wireless stations, quarters for officers and men, searchlights, oil-storage tanks, coal bunkers, magazines, fire equipment, guard-rooms, signal stations, hospitals, pay offices, dry docks, intelligence centres and all the vitally necessary stores, machinery and equipment of small dockyards.

To do this in the shortest possible time, and to maintain the supplies of such rapidly consumed materials as oil fuel, coal, food, paint, rope and shells for perhaps a hundred ships for an indefinite number of years, it was often necessary to lay down metals and sidings to connect the base with the nearest railway system. At many bases secure moorings had also to be laid by divers, and the channels and fair-ways dredged. The larger bases also required temporary shore defences, and booms arranged across the harbour entrances to prevent hostile under-water attacks.

Then came the problem of finding the personnel. The ships had already been provided for, but to keep them in fighting condition, and for the work of administration, it was necessary to have a shore navy behind the sea-going units. An admiral from the active or retired list was appointed to each base as the "Senior Naval Officer." Then came additions to his staff in the persons of executive and engineer commanders, officers of the Reserve, chaplains, surgeons and paymasters. With these departmental chiefs came their respective staffs of warrant officers, petty officers, wireless operators, engine-room artificers, motor mechanics, shipwrights, carpenters, smiths, naval police, signalmen, storekeepers, sick berth attendants and parties of seamen. Finally, a generous supply of printed forms and trainloads of stores.

This then, in brief outline, was the material which went to form the war bases of the auxiliary, or anti-submarine, fleets. In many cases much more was required, especially at such important depots as Dover, Granton and Queenstown. About the permanent dockyards, like Portsmouth, Devonport and Rosyth, or the Grand Fleet bases, nothing need be said here, because they do not come within the scope of this book. The same may also be said of that desolate but wonderful natural anchorage, Scapa Flow, the headquarters of the Grand Fleet in the misty north. Each of these mammoth naval bases had an auxiliary base for anti-submarine and minesweeping divisions.

With a knowledge of these essentials a more detailed description of a typical war base and the work of its staff may prove of interest. Taking as an example a large depot, supplying all the needs of over a hundred erstwhile warships, and situated in the centre of the danger zone, we find a central stone pier on which has been erected a perfect maze of wood and corrugated iron buildings, with the tall antenna of a wireless station, a little look-out tower and a gigantic signal mast from which a line of coloured flags is aflutter in the sea breeze. The shore end of the pier is shut off from prying eyes by a lofty wooden palisade with big gates, in one of which is a small wicket. Outside a sentry with fixed bayonet paces to and fro.

The first person inside the sacred precincts to greet the stranger is a keen-eyed "Petty Officer of the Guard." When the credentials have been examined the visitor is sent under the guidance of a bluejacket to the "Officer of the Day," whose

AFTER-DECK OF THE *HYDERABAD* SHOWING
QUICK-FIRING GUN ON DISAPPEARING PLATFORM.

ILLUSTRATION: AFTER-DECK OF THE *HYDERA-
BAD* SHOWING GUN RAISED TO FIRING POSITION.

"cabin" is inside the maze of corrugated iron and weatherboard. The doors flanking the passages traversed display cryptic lettering, such as I.O. (Intelligence Office), S.R. (Signal Room), S.N.O. (Senior Naval Officer), "Commander" (usually the second in command of the base), P.M.S.O. (Port Minesweeping Officer), C.B.O. (Confidential Book Office), M.L.Com. (Motor Launch Commander), O.O.W. (Officer of the Watch), "Officers only" (the wardroom and gunroom combined), and, finally, the O.O.D., or the abode of that much-worried individual, the Officer of the Day, whose duties happily terminate when his twenty-four hours of administrative responsibility are over, only, however, to return in strict rotation.

Again comes an apologetic examination of credentials, possibly followed by a few minutes with the admiral commanding,

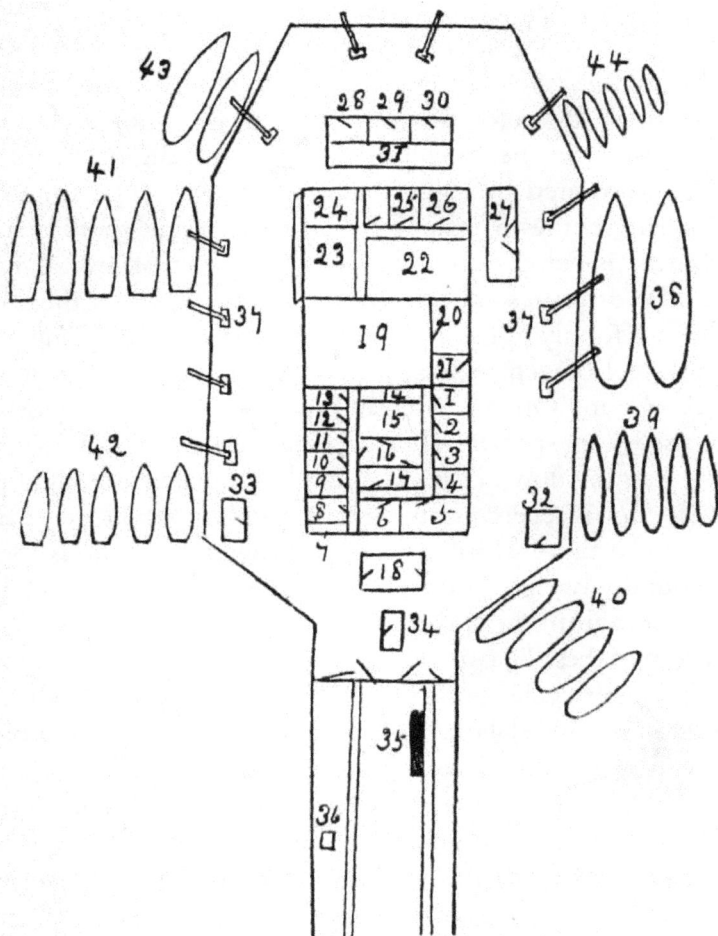

FIG. 17.

The central pier of a typical anti-submarine naval base. 1. Wardroom. 2. Sec. to senior naval officer. 3. Admiral's cabin (S.N.O.). 4. Flag commander (or lieutenant). 5. Base intelligence office. 6. Base commander. 7. Chaplain and gift store. 8. Drafting officer. 9. Store officer. 10. Chart-issuing office. 11. Cabin of the officer of the day. 12. Telephone exchange. 13. Warrant officers. 14. Pay office. 15. Fleet paymaster. 16. Paymasters and asst.-paymasters. 17. Writers and W.R.N.S. 18. Engineer-commander's office. 19. Men's quarters (for base duties and reserve). 20. Men's recreation room. 21. Petty officers. 22. Men's mess-room and adjoining galley. 23. Sick-bay. 24. Fleet surgeon. 25. Baths. 26. Baths. 27. Stores. 28. Boom defence office. 29. King's harbour master. 30. Hull defects office. 31. Police and cells. 32. Coaling office. 33. Wireless cabin. 34. Guard room. 35. Railway platform. 36. Sentry box. 37. Cranes. 38. Berths for armed yachts in harbour. 39. Motor launches in harbour. 40. Drifters. 41. Patrol trawlers. 42. Minesweepers. 43. Whalers. 44. Coastal motor boats. Larger ships, such as sloops, destroyers, "P" boats, coaling and ammunition hulks, lying out in basin.

and then the grand tour commences. First come the ships lying alongside the stone pier, with their short funnels belching black and very sooty smoke. These are the "stand-off" units, whose crews are enjoying a brief few hours ashore after days or weeks out on the dangerous seas beyond. Big drums of oil are being lowered by ropes on to their decks. The sound of hammering comes from more than one engine-room, where machinery is being overhauled. On the decks of several, men with little or no resemblance to the clean "Jacks" of the naval review are fondly polishing, painting or greasing the long grey barrels, steel breech mechanism, or the yellow metal training wheels of guns. Others are cleaning rifles, which have recently been used with special bullets for sinking floating mines. One ship is washing down decks after coming in late from night patrol; another is receiving its three-monthly coat of grey paint; while on to the deck of a whaler—black and ominous-looking—hundredweights of provisions in boxes and bags are being lowered from the quay.

Astern of these lie two tiers of light grey spick and span motor launches, their decks spotlessly white, and their small canvas and glass screened wheel-houses ill concealing polished brass indicators, Morse signalling key, electric switches, binnacles and other paraphernalia. Behind these lie the 40-knot coastal motor boats, like miniature submarines, with torpedoes in cavities in their aft decks, and little glass-sheltered steering-wells. Further towards the head of the pier is a line of big flat Scotch motor drifters, built for rough weather with 9-inch timbers, their decks a maze of wire nets, glass floats and brick-red chemical canisters.

On the opposite side of the pier, in front of the S.N.O.'s cabin, lies a big grey yacht with four 12-pounder guns and an anti-aircraft weapon pointing over the sky-reflecting water. Lying out in the basin are big minesweepers, looking more like pre-war third-class cruisers, two slim-looking dark grey destroyers, a dredger and a few nondescript craft.

Inside the first row of iron sheds are stores, with barrels of tar, drums of paint, immense coils of rope and a naval "William Whiteley's"—in which anything from a looking-glass to a ball of string, or a razor to a dish-cloth, can be obtained in exchange for a signed form from the Naval Store Officer, whose cabin near by is a maze of similar forms of all colours.

Then a worried-looking man hurries by and the O.O.D. smiles. "He's the coaling officer, and there's some twenty ships waiting to get alongside to take the beastly stuff aboard," is the laconic explanation.

A cabin marked I.O. is entered—every room is a cabin in a naval base. Here the walls are decorated with innumerable charts with mysterious red lines. A curious device, with the names of all the ships belonging to the base painted on wooden slides, reaches across one side. It is the indicator which shows at a glance the ships at sea and those in harbour, the names of those under repair, the unit to which each vessel belongs and when she goes out or comes in for "stand-off."

This is the Intelligence Office, and signals and wireless messages from the patrols and battle fleets are being almost continuously brought in and carried out by messengers. The Commanding Officer (C.O.) of a minesweeper is making inquiries about tides and the exact position on the chart of a newly located mine-field. Another officer is locking a black patent-leather dispatch-case— he is the King's Messenger or, more correctly, the "Admiralty Dispatch Bearer," who carries to and from London and the fleets all the secret correspondence and memoranda of the Naval War Staff and other important departments. A big safe in the corner of the cabin contains the secret codes and ciphers used when transmitting messages, and two overworked officers are busy at near-by desks translating signals to and from "plain English."

The next cabin contains the admiral's secretary and his staff of writers. Here a flotilla commander is receiving his "sailing orders," without which no ship proceeds on a voyage. Adjoining this is the Pay Office, in which, with the exception of a newly joined recruit mortgaging his pay for two weeks ahead— he knows that he will be at sea for that time—there is a decided air of quietude. The rush in this abode of paymasters comes at the end of each month, when all the officers arrive in a body to demand the meagre fruits of their labours.

Sandwiched between the clean and varnished cabin of the Base Commander, who is "taking" defaulters, and the camp-bedded apartment of the O.O.W. is a most interesting little combined cabin and store, presided over by the Chaplain. Here are piles of woollen socks, cardigans, balaclavas, mitts and other clothes knitted by the thoughtful women of the Empire for their sailor sons. Here seamen are estimating the cold-resisting qualities of different garments—for winter in the North Sea is the next thing to Arctic exploration. Officers are popping in and out to borrow a pile of books—thrice blessed were the senders of these donations. The corner of the cabin is piled with fresh vegetables, but alas! the cry is apples! No exhortations to righteousness adorn the walls, and the chaplain is joking with a big stoker who is distractedly turning over the cardigans in search for one large enough to encompass his massive frame. A signal boy slips in, gets chocolate, gives a breathless thanks and slips out just in time to avoid the playfully raised hand of the P.O. of his ship. Two deck hands, covered in coal dust, put their heads round the door to ask if they can have a bath, and the indefatigable chaplain hands them the keys of the room provided for the purpose by the generous.

Religion here is more practical than theoretical. If a man swears when the "Padre" is present he pays a small fine, which goes to the recreation or other needy fund. The Commander is not immune from this law at the base under review, and has more than once been "heavily fined" for giving his true opinion of German sailors and winter weather.

The next cabin is that of the O.O.W., a seething mass of officers demanding "duty boats" and pinnaces to convey them to and from their ships lying out in the fair-way. Others are expostulating about being ordered to sea during their "stand-off," informing everyone what a rotten service the navy is, crossing-sweeping is a sinecure compared with it. Then a few pass on to the cabin near the men's quarters. Here the "Drafting Officer" is trying to palm off a deck hand on the C.O. of a trawler, who is vainly explaining that he must have a signalman. A telephone rings and news comes from the "Sick Bay" that an engineer has been badly burned and will be unable to go to sea with his ship. The distracted drafting officer searches through his lists of reserves for some competent man to take the place of the casualty.

Peace reigns in the adjoining department, where a grey-haired veteran is issuing charts, "Sailing Directions," "Tide Tables" and "Warnings to Mariners." In the near-by engineer-commander's office worried experts are wrestling with innumerable problems relating to M.L. motors, steam capstans, steam steering gear, electric dynamos, damaged propellers, broken shafts, boiler cleaning and the numerous imperfections of overworked ships' engines.

The Boom Defence staff is placidly serene. The turn of this department comes after a heavy gale has damaged the submarine nets, chains and buoys. The torpedo officers and their "parties" are discussing the best way of moving four of these steel monsters from a neighbouring depot ship to a new "Q" boat with only a rowing-boat at their disposal—soon the O.O.W. will be called upon to supply a drifter for the purpose.

In the ordnance store a veteran P.O. is trying to make his list of returned brass shell-cases correspond with the number of shells supplied to various ships six months before. He knows the sailors' fondness for shell-cases as ornaments in their little faraway homes, and, failing to make all the figures agree, decides that some *must* have been "washed overboard."

The Port Minesweeping Officer is discussing with his sea commanders the clearing of a new mine-field laid by U-C-boats within the past few days, when a sudden stir is caused by the arrival of a signal from the wireless room to the effect that one of his vessels has struck a mine in lat.—long.—and is sinking. He appeals by telephone to the M.L. commander and in less than ten minutes a flotilla of fast launches is racing at 19 knots to the rescue.

In the Admiral's cabin there is to be a conference of senior officers later in the day to decide on the best means of ridding the seas within that area—and each base has its own area of sea—of a hostile submarine which has been inflicting undue loss upon shipping, its latest victim being a Danish barque.

The combined wardroom and gunroom has some twenty occupants, reading the newspapers and magazines, warming

MOCK-WHEEL AND COMPASS-PEDESTAL OF THE *HYDERABAD*

WHICH COLLAPSE AND LEAVE A CLEAR RANGE FOR THE GUNS

themselves before the two big fires, or talking in little groups. This base has suffered some heavy losses lately, but reference to those "gone aloft" is seldom made, except quietly and a little awkwardly. The talk is of theatres in neighbouring towns, the respective merits of certain types of ships and weapons, the prospects of early leave, the dirty warfare of "Fritz" or the "beauties" of the North Sea in winter.

In this room all questions of rank and precedence are more or less waived. There are, of course, differences, especially when the wardroom, or abode of senior officers, does duty also as a gunroom for the juniors. But here there is camaraderie and an absence of iron discipline, although a sub-lieutenant would be extremely ill advised either to drop the prefix "Sir" or to slap the Commander on the back in an excess of joviality, relying on "neutral territory" to save him from rebuke. It is, however, no uncommon event to see all ranks of officers engaged in a heated debate, or groups of juniors laughing round the fire while their elders are vainly trying to concentrate their minds on the latest Press dispatches. Games are played and glasses clink merrily, but in a gunroom there is a very strict limit as to both time and quantity, though none regarding volume or discordance of sound.

****

Passing on to the organisation of the flotillas for sea, we find in this large base six minesweeping units, two being composed of fast paddle sweepers and four of trawlers. The former are used for distant operations and comprise nine vessels. They work in pairs, but the extra ship is available to sink mines cut up by the sweeps of the others, and to be immediately ready to beat off submarine attacks.

The trawlers are engaged in sweeping *daily* the approaches to the harbour and a recognised channel up and down the coast. Their work overlaps with that done by the ships belonging to the neighbouring bases. In this way the "war channel," about which more will be said later, was kept free of mines, and afforded a safe route for ships from the Thames to the Tyne, and in reality to the northernmost limit of Scotland.

This important duty was seldom left unperformed even for a day, except during fierce gales. Often the discovery of a distant mine-field caused many ships to be concentrated on clearing it, and the number available for the "routine sweeps" was consequently reduced, but longer hours of this arduous and dangerous work made up the difference, and the work went on in summer fog and winter snow for over four years.

The anti-submarine patrols were composed of five ships each, under the command of the senior officer of the unit—frequently a lieutenant with the responsibility of a captain. Their work lay out on the wastes of sea lying between England and Germany. It was seldom that the whole five vessels of each unit cruised together, the usual method being to scatter over the different "beats" and rendezvous in a given latitude and longitude at a specified time and date. They were usually able to communicate with each other and with the base on important matters by wireless. Their periods at sea varied from ten days to three weeks, with a four days' "stand off" when they came into harbour. But of this time one day at least was spent in coaling and provisioning the ship ready for the next patrol. This ceaseless vigilance on the grey-green seas of England's frontier was seldom interrupted for more than a few days in the year by impossible gales. Anything short of literally mountainous seas did not prevent the trawler patrols from riding out the storm carefully battened down and with just sufficient speed to keep head to sea.

The drifters were divided into patrol units, boom defence flotillas and under-water or mine-net units. Their work was thus more varied but equally as arduous and risky, as the loss of 30 per cent. of the entire fleet of over 1000 ships affords undeniable proof. The periods of sea duty were similar to those of the trawlers.

The motor launches at each base had some hundred square miles of sea to guard, and hunted in fives. The rough weather these plucky little ships endured in the open sea in mid-winter, the intense cold—for there was no proper heating appliance—and the state of perpetual wetness made their duties among the most arduous in the sea war. Later pages of true narrative will show to the full the work of these gnats of the sea.

In addition to all these flotillas there were convoy ships, whaler patrols, "Q" boats and a number of special duty ships. The work of the former was of the most exacting character, and left the crews of these vessels but little time ashore. In the base under review so arduous were the duties of the convoy ships that it became a matter of self-congratulation for patrol and sweeper officers and men that their ships were not so employed, and this by men who sailed submarine and mine infested seas for an average of 270 days in each year!

It must not be assumed that when in harbour there were no duties to be performed by either officers or men of sea-going ships. They had, on the contrary, to furnish anchor watches, shore sentries, duty crews for emergency pickets, prisoner guards, working and church parties, to attend drills, rifle practice, gun practice and instructional parades. The officers had similar shore duties to perform, which left them little time to rest from the strain of keeping watch and ward on the death-strewn seas.

CHAPTER 9

The Convoy System

Although the convoy system was employed at the beginning of the war for the transport of the Imperial armies to France, and subsequently for all the Allied troop movements overseas, it was some three years later before it was extended to the entire British Mercantile navy, on which the United Kingdom depended for too many of the necessities of civilised life.

The rapid development of submarine piracy, however, compelled the Admiralty, early in the year 1917, to resort to what was merely a new form of the old system of protecting seaborne trade. This comprised the collection of all merchant ships passing through the danger zones into nondescript fleets, and the provision of light cruisers, destroyers, torpedo-boats, trawlers and occasionally (for coastal convoys) of patrol launches to escort them. Certain types of aircraft were also frequently used for observation and scouting purposes.

Previous to the adoption of the convoy system a merchantman, whether it was a fast-moving liner or a sturdy but slow ocean tramp, *zigzagged* through the danger zones with lights out and life-boats ready. Many were the exciting runs made in this way, with shells ploughing up the water around and torpedoes avoided only by the quick use of the helm; but the courage of our merchant seamen was of that indomitable character exhibited now for over three centuries, since the days of Drake, Hawkins, Raleigh and the other sea-dogs of old.

But the danger zones grew wider as the radius of action of newer and larger German submarines increased. At last no waters were immune, from the Arctic circle to the Equator, or from Heligoland to New York.

The hour was one of extreme peril for the sea-divided Empire. To lose several hundred ships, with many thousands of lives and much-needed cargoes of food and munitions, when the valiant armies of civilisation were battling with the Teuton hordes, was bad enough; but if the enemy had been able, by casting aside the laws of humanity and sea war, to compel British ships to remain in harbour or meet certain destruction on the high seas, the result could only have been the complete failure of the Allied cause, the conquest of Europe and the fall of the greatest political edifice since Imperial Rome.

Between the world and these catastrophes, however, stood the undefeated Mercantile Marine and the Allied navies. Councils were held in the historic rooms of Whitehall and the old convoy system emerged from the archives of Nelson's day. The commerce raiders were no longer the canvas-pressed privateers of the sixteenth, seventeenth and early eighteenth centuries, who fought a clean fight, often against great odds, but were submarine pirates of the mechanical age, who only appeared from the sea depths when their victims had been placed *hors de combat*.

It is an old axiom of war that new weapons of attack are invariably met by new methods of defence. So it was with the convoy system which gave the death-blow to German hopes of a submarine victory. In order to understand this *new* method it is necessary to study the accompanying diagram, which, however simple it may appear on paper, is extremely difficult to carry out in practice. At each great port there was a convoy officer, who assembled the merchant ships when they had been loaded and explained to their captains the exact position each ship was to occupy when the fleet was at sea. Printed instructions were handed round urging each vessel to keep its correct station, stating the procedure to be adopted in the event of an engine breakdown, giving the manoeuvres which were instantly to be carried into effect when an attack was threatened, and finally the special signals arranged for communication between the merchantmen and their escort by day and by night.

The number of vessels composing a convoy varied, but often exceeded twenty big cargo ships, carrying some 120,000 tons of merchandise, or six liners, with 20,000 troops on board, while the escorting flotilla consisted of a light cruiser, acting as flagship, six destroyers, two special vessels ("P" boats) towing observation airships, and some eight or ten trawlers, with possibly one or more seaplanes and several M.L.'s for the first few miles of the voyage. The destroyers were spread out ahead and on the flanks of the fleet, and by using their greatly superior speed were able to zigzag and circle round the whole convoy.

In the event of an attack the whole fleet turned off from the course they were steering at a sharp angle, showing only their

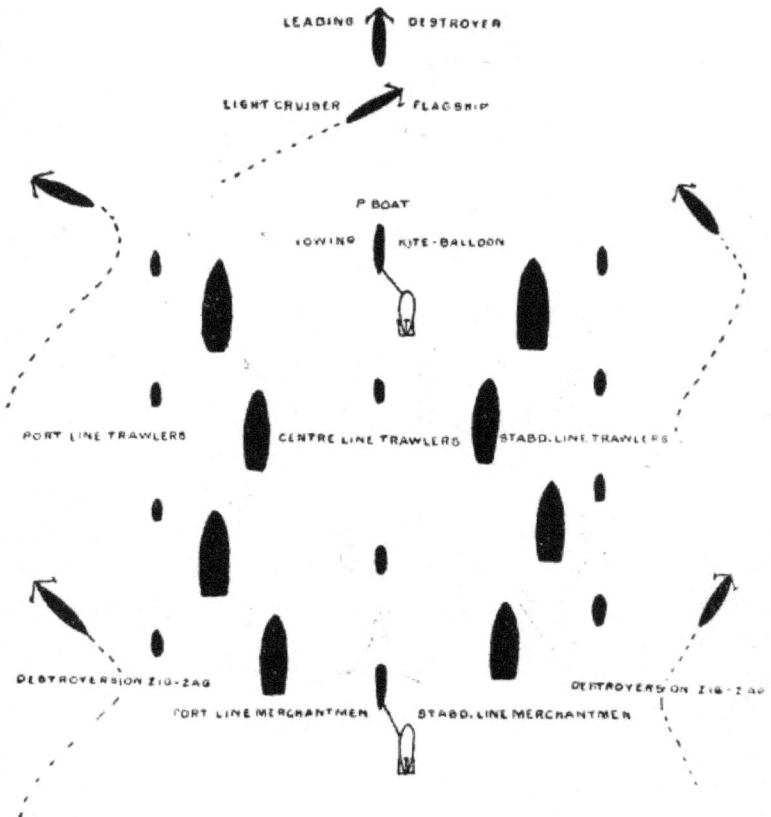

FIG. 18.
Diagram showing the disposition of a convoy of troops, munitions or food.

sterns to the U-boat. A destroyer acted as rearguard to prevent any of the convoyed ships from straggling. When the fleet had arrived at a rendezvous far out in the open sea, where the danger of a submarine attack was much less, the escort handed over their charges to one or two ocean-going cruisers, which stayed with the merchant ships throughout the remainder of their voyage.

The escorting flotilla then cruised about in the vicinity of the rendezvous until an incoming convoy appeared. These ships were then taken over from their mid-ocean cruiser guard and escorted back through the danger zone to port, and so the game of war continued until months became years.

All this may sound straightforward and quite simple, but there were difficulties, to say nothing of dangers, which made it a most arduous operation. First came the speed problem. Every merchant ship differed in this important respect, so the speed of the slowest unit became the speed of the entire fleet, and this reduction made an attack by under-water craft much easier of accomplishment. Hence the call for "standard ships," which is a point that should be borne in mind by future generations as a safeguard against blockade. Then came the question of destination, which increased the number of escorting flotillas, and

A MOTOR LAUNCH CLEARED FOR ACTION

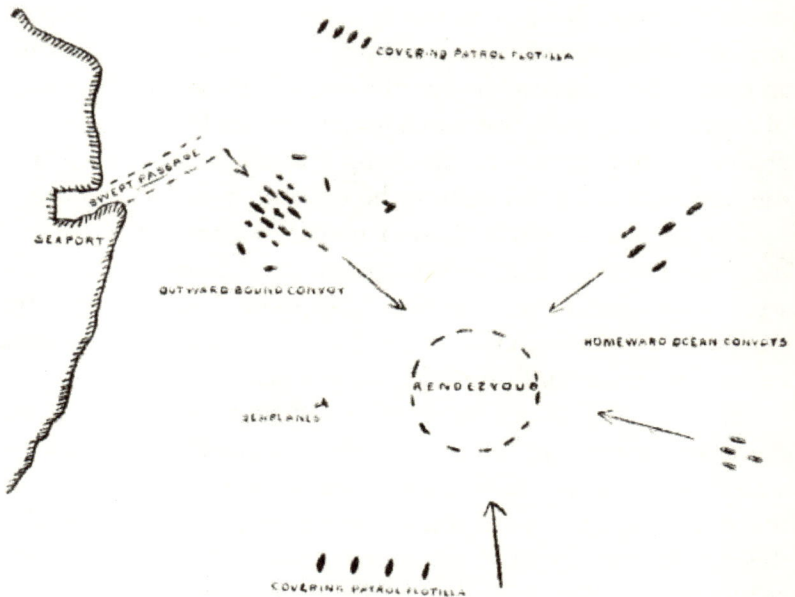

FIG. 19.
Diagram showing the convoy system

especially ocean cruiser guards, required for a given number of cargo ships. Next there was the loading and unloading to be considered, involving long hours and hard work by the men on the quaysides. This great difficulty was one of the reasons for the formation of docker battalions. Coaling such big fleets by given times caused many grey hairs to appear where otherwise they would not have been. Finally there was the danger of mines having been laid in the fair-ways leading to the port, which necessitated every convoy being met by special vessels to sweep the seas in front of each incoming and outgoing fleet.

All this and more had to be contended with and overcome before each convoy was able to sail. Then danger and difficulty came hand-in-hand. On a bright morning, with probably a fresh breeze blowing and a choppy sea, the work of the escorting flotilla was easy, but with such climatic conditions the risk of attack was so great in the waters around the coasts that troopships usually left harbour under cover of night. No lights were then allowed, and it will not be difficult for readers to imagine what it meant to be pounding through a black void in a fast-moving destroyer, against, possibly, a heavy head sea, with some twenty or thirty big ships in the darkness and spray around. Thick sea-mists were the cause of endless trouble, for the safety of an invisible fleet depended on the vigilance of a half-blind escort. Winter gales scattered the ships and rendered signals invisible. Attacks came from the most unexpected quarters and often from more than one point of the compass at the same time. However, relief came at last, on that never-to-be-forgotten morning when Sir David Beatty and his admirals accepted the unconditional surrender of the German fleet and its unsunk submarines.

Were this chapter to end with the foregoing description of the convoy system the reader would not be in possession of the full facts from which to gauge the importance of the work. Something must be said of what was accomplished. First in order of importance came the transport of many millions of soldiers not only from England to France, but also to and from every colony and dominion of the world-wide Empire. By August, 1915, the British navy had transported, across seas infested with submarines and mines, a million men without the loss of a single life or a single troopship.[5] The first Canadian army of 33,000 men crossed the Atlantic in one big fleet of forty liners, under the escort of four cruisers and a battleship, in October, 1914, without accident. Transports to the number of 60 were required to convey the first Australian army over the 14,000 miles of sea to Europe, and it was while convoying this huge fleet that the cruiser *Sydney* chased and destroyed the German raider *Emden*. The Russian force which rendered valuable service in France was safely convoyed over the 9000 miles of sea from Dalny to Marseilles. Never once during the four and a half years of war was the supply of food, munitions and reinforcements, or the return of the wounded—to and from the many theatres of land operations—seriously hindered by the German, Austrian or Turkish navies.

---

5. When writing of the navy in this connection due praise should be given to the Mercantile Marine, which this war has proved to be a very important part of the *true* sea power of Great Britain.

Turning to the gigantic task of guarding England's food supply, we find, in one notable case, an example of the good work performed almost daily for nearly five years. Over 4500 merchant ships had been escorted across the North Sea to Scandinavian ports alone before the disaster of 14th October 1917 befell the convoy on that route. On that occasion the anti-submarine escort of three destroyers were intercepted, midway between the Shetland Islands and Norway, by two heavily armed German cruisers. The destroyers fought to the last to save their charges, but unfortunately only three merchant ships succeeded in getting safely away. Five Norwegian ships, three Swedish and one Danish ship were sunk. From this it will be observed that not only British merchantmen were protected by escorts.

The second attack on the Scandinavian convoy occurred on 12th December. The escort consisted of two destroyers, the *Partridge* and *Pellew*, with four armed trawlers. Fortunately the convoy was comparatively a small one, for it was attacked and almost totally destroyed in the North Sea by four of the largest German destroyers. H.M.S. *Pellew*, although badly damaged, succeeded in returning to England.

It may be rightly thought that in both these cases the escorting flotilla was not strong enough, but it should be remembered that if heavier ships had been employed they would have been much less able to cope with a submarine attack. The escort in both cases was purely an anti-submarine defence, and only on the Scandinavian and Netherlands routes was a surface attack at all possible, because all exits from the North Sea were securely closed by the strategic positions occupied by the Grand Fleet and the battle cruiser squadrons, in conjunction with subsidiary fleets at Harwich and extensive mine-fields.

When it became apparent that surface as well as submarine attacks on the North Sea convoys had to be provided against, other means were promptly adopted, and no further disasters occurred.

The strong escort accompanying the transports bringing to Europe the first American army were attacked at night by a submarine, but succeeded in avoiding the torpedoes fired. This was due to the smartness with which the United States warships were manoeuvred. Three subsequent attacks on the same convoy route also failed.

The Report of the War Cabinet for the year 1917 gives some remarkable figures in support of the convoy system. On the Atlantic routes about 90 per cent. of the ships were formed into fleets and escorted. From the inauguration of this system the loss on these routes from all causes was 0.82 per cent., and if all the trade routes to and from the United Kingdom are included, the loss was only 0.58 per cent. With these figures in mind, who will deny that the navy is the surest form of national as well as Imperial insurance?

CHAPTER 10

The Mysteries of Submarine
Hunting Explained

When all is said and done, anti-submarine warfare is very like big-game hunting. Success depends entirely on the initiative, skill and resource of the individual hunter. Contrary to general belief, there is, at present, no sovereign remedy for the depredations of under-water craft with their torpedoes and mines. There are, however, several recognised methods of attack and defence employed by surface ships in this newest form of naval warfare.

When the new navy took the seas in 1914-1915, bases were established not only round the coasts of the British Isles, but also in the more distant seas. The principal danger zones were, however, the North Sea, the English Channel, the Irish Sea, the Mediterranean and the eastern portion of the North Atlantic. It was through these waters that every hostile submarine must pass on its voyage out and home.

This geographical factor restricted the theatre of major operations to some 180,000 square miles of sea. Minor offensive measures might have to be adopted against individual U-boats cruising at long distances from their bases, as actually occurred off the United States coast, but the fact of Germany possessing large submarine bases only along her own North Sea coast, and temporary ones on the Flanders littoral, enabled a concentration of Allied anti-submarine craft to be made in the narrow seas which afforded the only means of entry and exit to and from those bases.

The same may be said of Austria in the Adriatic and of Turkey behind the Dardanelles.

This favourable combination of circumstances would not occur if (however unthinkable) England, France or the United States were ever to wage a rigorous war against shipping. The large number of overseas naval bases possessed by these Powers would cause every sea to become a danger zone within a few hours of the commencement of hostilities. No effective concentration of hostile surface craft would be possible with the zone of operations spread over the water surface of the entire globe, and if the bases themselves were secured by predominant battle fleets, or numbers of heavily armed monitors, the seas would quickly become impossible for purposes of hostile transport.

This geographical restriction of the German and Austrian danger zones made effective concentration of the Allied anti-submarine fleets and their devices possible. The 180,000 square miles of sea, forming the theatre of major operations, was, on special charts, divided into areas, comprising a few hundred square miles of sea. Each area was given a distinctive number, and a base was established for its own patrol and minesweeping fleet.

The areas themselves were again subdivided on special charts into squares or sections. Each square covered a few leagues of sea and was known by an alphabetical sign. In this way the waters of the submarine danger zone were divided into areas, with their bases and protective fleets, and squares with their respective squadrons or ships.

Each square of sea was covered once or twice daily by its own patrol ship or flotilla. Where the danger was less the patrol was not so frequent and the squares were almost indefinite in size, but where the chances of successful operations were exceptional, as in the Straits of Dover, additional offensive measures were employed (see under Mine Barrages).

This, then, was the chess-board on which the game of submarine warfare was played. To facilitate communication between the different patrols spread over the squares of sea, wireless was fitted in many ships, and war signal stations were erected on prominent points of land. Each base was able to communicate by wireless with any of its ships out on patrol duty, and was also connected by land-line telegraph, telephone and wireless with *naval centres*.

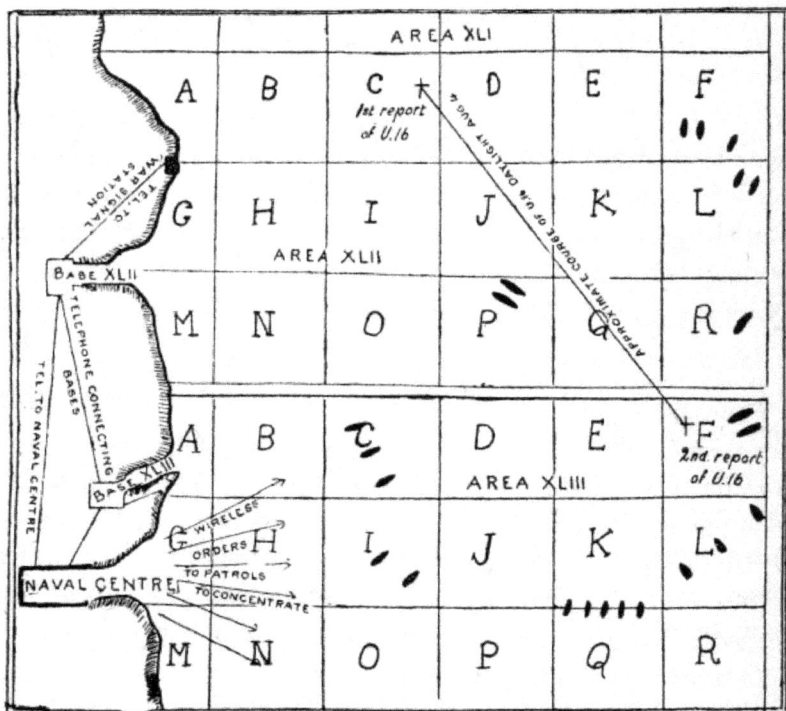

Fig. 20.
Diagram showing division of sea into anti-submarine patrol areas.

These latter were head intelligence offices, usually situated at the great bases of the battle fleets. In this way any concentration of hostile surface warships noticed by the patrols (sometimes submarines were employed, especially in the Heligoland Bight) could be communicated in a few minutes to the admirals commanding the Grand Fleet, the Battle Cruiser Squadron or other large fighting organisations.

At the naval centres the movements of hostile submarines were recorded on charts. If, for example, it was reported from a patrol boat that the U16 had torpedoed a ship in square "C," area 41, at 10 a.m. (G.M.T.[6]) on 4th August, and the patrol had arrived on the scene too late to be of any service, a warning could be wirelessed to hundreds of vessels on the seas surrounding the scene of outrage to keep a careful look-out for the U16.

6. Greenwich mean time.

Subsequently a further message might come to the naval centre that the same submarine had been chasing a merchantman in square "D," "E" or "F" in the adjoining area. A concentration of fast ships, such as destroyers, M.L.'s or coastal motor boats, could then be made so as to intercept the raider or enclose her in a circle while other vessels hunted her down.

In a like manner important convoys coming down the coast, or entering a danger zone from the open sea, could be met by a local flotilla and escorted to a rendezvous with a flotilla from the adjoining area. In this way they were passed through the submarine and mine infested seas to and from their harbour terminus.

Almost the same methods were employed in dealing with the thousands of German mines. But to describe that part of anti-submarine warfare here would be to encroach on the subject of a succeeding chapter.

PATROLS

The *method* of patrolling the areas and squares of sea was comparatively simple, though the same cannot be said of the actual work. The lines of patrol were called "beats," and there was usually an "inner" and an "outer" beat for each unit or flotilla of ships. If when a ship (or a unit) reached her allotted square, from which the line of patrol extended, she elected to proceed on the *inner beat*, she would generally accomplish the return journey to the point of departure on the *outer beat*, thus covering her respective zone of patrol, but leaving the exact route to the discretion of the commanding officer. In this way no hostile submarine with a knowledge of the system could be sure of when or where a patrol ship would be met. In the same way it was left to the commander of a flotilla to either divide his ships into pairs, single units, or to maintain them as a homogeneous fleet, so that any combination of hostile submarines could not be made which would be sure of being able to attack a *single* patrol. Such an enemy combination might encounter a single ship, but it might also walk into the arms of a whole flotilla; or it might attack a single ship only to find itself surrounded by a following fleet.

The beats which were most distant from the base were given

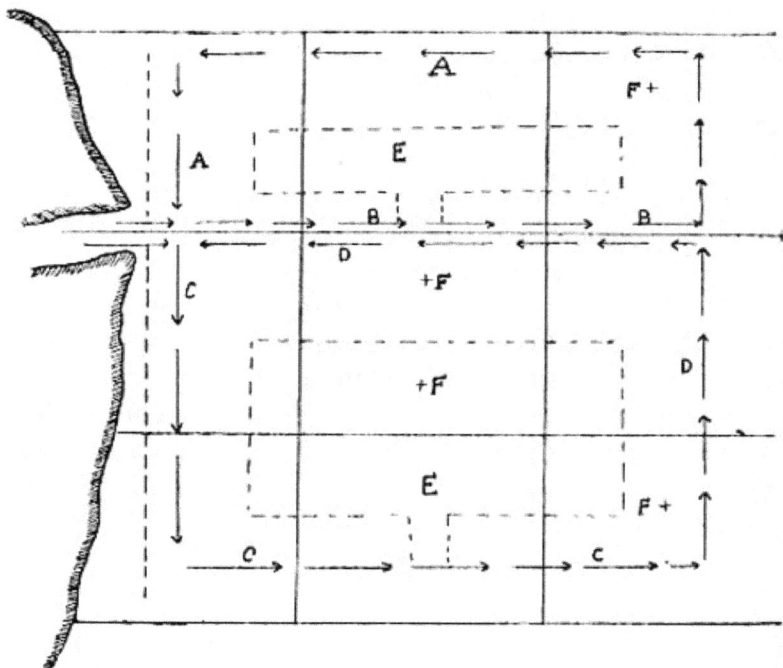

FIG. 21.

Diagram showing how an area is covered by patrols. *A*. Unit or flotilla of ships may proceed out from the base on course indicated by arrows *B*, which would be called the "Northern Inner Beat," and return to harbour on course *A*, "Northern Outer Beat." Other units of ships would simultaneously follow the course *E*. These and adjacent squares of sea would be covered daily by one or more ships of each unit. The southern half of the area would be patrolled in the same way. The "Outer Beat" is shown by the arrows *C*, and the "Inner Beat" by the arrows *D*. The points +*F* show the possible positions of armed patrols acting independently of any unit or flotilla.

to the largest ships. This was done because it was often impossible for the more distant patrols to reach a place of shelter before one of the fierce gales which swept the northern seas was upon them. Trawlers, large steam yachts and converted merchantmen were usually employed on squares more than one hundred miles distant from a harbour of refuge, while motor launches kept watch and ward on the seas closer inshore.

The duration of patrols varied according to their position. Some lasted three weeks and others only a few days or hours. When the ships returned to their base after a spell at sea they were given a corresponding "rest" in harbour. A three weeks' patrol meant several days' "stand-off," while a two or three days' patrol entitled the ship to twenty-four hours in the comparative comfort of a harbour.

It must not be imagined, however, that a stand-off meant entire idleness or thorough rest. There were duties to perform which robbed it of much that it was intended to give. Ships had to be coaled, provisioned, painted or repaired. Engines had to be overhauled, sentries posted ashore, a guard to be furnished, and every day one ship in each unit that was in harbour had to be manned and in readiness for emergencies.

HYDROPHONE FLOTILLAS

We now come to the actual methods employed by surface craft when attacking submarines. Although, as previously stated, much was left to individual initiative, there were, nevertheless, certain recognised methods.

Taking as an example the operations of a hydrophone flotilla of armed motor launches, the number of vessels forming the unit was usually five. When out scouting for the enemy they proceeded in line-abreast for about one sea mile, then stopped their engines and listened on their hydrophones for the noise of a submarine cruising in the vicinity. If nothing was heard the mile-long line of miniature warships advanced another mile and again stopped to listen. This manoeuvre was repeated until one or other of the ships heard the familiar sound of a U-boat. Nothing might be visible on the surface of the sea, but if this was the case and the noise came up from the ocean depths over the electrified wires of the detector, it was conclusive proof that a submarine was in the near vicinity.

The M.L. first detecting the noise hoisted a signal (flag by day and coloured electric light by night), giving the direction from which the sound came (see Fig. 22). The next ship in the line to receive the sound on its instruments then hoisted a signal, also giving the bearing—*i.e.* N.N.W., E.S.E., etc. If the two coincided in regard to direction, the attack commenced. If, however, they did not agree in this important respect, the line of patrol ships advanced another mile and listened again.

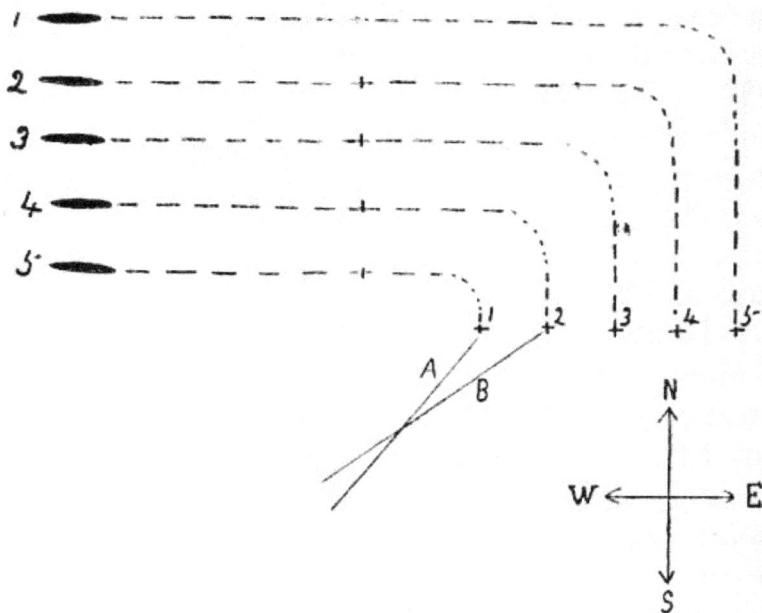

FIG. 22.

Diagram illustrating the operations of a hydrophone flotilla composed of armed motor launches. Each vessel is given a number, and the flotilla proceeds in line-abreast along the course shown by the dotted lines. Each vessel is one mile from the other, and the whole line stops by signal at the point marked with a cross. Hydrophones are put in operation, and after a period of listening the flotilla continues on its course, as no submarine sounds are heard. The flotilla turns to head south, and a stop is again made to listen on the hydrophones. This time the sound of a hostile submarine is heard by vessel No. 1, bearing S.W. This report is confirmed by vessel No. 2 hearing the same sound, bearing a few degrees farther W. The two bearings *A* and *B* are then drawn on a chart, and the point where the two lines cross is the approximate position of the invisible submarine. The attack with depth charges is then ordered.

The flag-ship of the unit on receiving the direction from one or more ships marked the lines of sound on a chart (as in Fig. 22), and when this was substantiated by another ship the point where the two lines crossed was known to be the position of the hostile submarine, and the attack was ordered.

As to the exact method of an anti-submarine attack little need be said here beyond the fact that the ships advanced at full speed, manoeuvring into a special formation which enabled them to cover about half a square mile of sea with the explosive force of their collective depth charges.

When the attack had been completed all vessels engaged resumed their stations and waited with quick-firing guns ready in case the monster should rise from the deep to make a dying effort to destroy her pursuers.

The tactical methods of anti-submarine attack were, of course, numerous, and they varied according to the speed of the surface ships engaged. What was possible of accomplishment by fast-moving coastal motor boats or the larger-sized M.L.'s proved impracticable for the more heavily armed but slow-moving trawlers and drifters. The tactics of these latter craft were often of the simplest character, and consisted principally of either independent attacks with the aid of hydrophones and depth charges, or, more frequently, the assumption of an innocent air in order to induce the submarine to open the attack at close range.

In many respects this proved the most effective method of anti-submarine warfare. Not only did it frequently cause the under-water craft to rise to the surface and commence the attack by gun-fire, in order not to expend a valuable torpedo on what appeared to be an unarmed and helpless ship, but it also produced a *moral* effect throughout the German submarine flotillas.

When a few U-boats had been either sunk or damaged in this way the news that every Allied ship was heavily armed circulated among the enemy personnel, and they became very nervous of attacking in any position except totally submerged. This meant the loss of at least one torpedo, out of from five to ten carried, for every attack made, whether successful or unsuccessful, and the latter were predominant.

It soon became apparent that either they must risk surface attacks and so save their torpedoes, or else curtail their cruises to meet the rapid expenditure of their only submarine weapon. This does not, of course, cover the activities of under-water mine-layers, whose nefarious purpose consisted simply of laying their mines wherever they appeared most likely to catch Allied shipping. These craft were usually armed with torpedoes as well as mines, to enable them to continue the work of destruction when the cargo of the latter had been safely laid. In this way the problem of combating the German submarine offensive re-

The result of a direct hit: a photograph left by the Germans in Ostend showing a coastal motor boat washed ashore after the great raid.

solved itself into two parts, one being to checkmate the commerce raider and the other the mine-layer. With the second of these difficulties we shall deal in a later chapter.

Many merchantmen, both Allied and neutral, owed their escape to this camouflage warfare, which was brought to a high pitch of perfection and daring in the now famous mystery ships.

What may be said to form the second method of anti-submarine warfare was the decoy or camouflage system. Of primary importance in this category were the mystery ships already described, but there were also innumerable other *ruses de guerre* which increased its efficiency.

To describe one of these will enable the reader to draw on his own imagination for the remainder. A vessel was steaming in from the Atlantic and was about a hundred miles from the Cornish coast when she was attacked by a submarine above water. The surface ship was heavily armed, but instead of using her weapons at once she sent out frantic wireless signals for assistance. Every few minutes the call went far and wide in plain Morse.

The shells from the submarine splashed into the sea around, but none struck the target for some minutes. Had the surface ship desired, she could in all probability have avoided the underwater craft by using her superior speed, but instead she dropped back, allowing the submarine to catch up to her, and the shells began to burst unpleasantly close.

Still the frantic wireless calls went forth. First the simple message: "I am being attacked by a large German submarine." Then the vehemence increased to: "I am being heavily shelled." A few minutes elapsed and then the call: "Help. Submarine gaining on me." And finally: "Abandoning ship."

At this point the submarine was close astern and the liner slowing down preparatory to lowering her life-boats. The shells were damaging her superstructure, but a heavy swell interfered with the German marksmanship. Then came the surprise. A life-boat on the liner's poop was hoisted clear of the deck and from under its cover there appeared the lean grey muzzle of a 4.7-inch gun. A few sharp blasts of cordite and the submarine sagged and disappeared.

The captain of the liner had noticed when first attacked that the submarine was fitted with wireless and the calls sent out by him were in *plain Morse code*. On the strength of these the German commander had saved his torpedoes but lost his ship.

****

Another form of anti-submarine tactics was the employment of indicator and mined nets around an apparently disabled ship, or in lines across narrow channels known to be used by German submarines on their way to and from their bases. This method has, however, received full mention in other chapters.

****

What may be termed the third system of anti-submarine warfare was the use of extensive mine barriers, specially laid to catch submarines attempting to pass through them under water. The surface of the sea was patrolled by shallow-draft vessels and the under-seas guarded by mines. If a submarine was sighted in the vicinity of one of the mine barriers already described she was attacked and forced to submerge herself in order to escape destruction from the guns of the pursuing surface flotilla. From that moment her fate was sealed. By cautious manoeuvring and using to full advantage their great superiority of speed (20-40 knots against 6-10 knots) the surface ships were able to head their quarry into the mine-field. Usually the submarine dived deep in order to throw her pursuers off the track, and all unconscious of the deep-laid mines in thousands she plunged to her doom—a heavy rumble, followed by an upheaval of the surface, and the chase was over.

This method, when carried out on the vast and scientifically sound principle described in a previous chapter, offers the best possible antidote to the submarine. Its employment in the Great European War placed the seal of complete success on the Allied anti-submarine offensive. It should, however, be remembered that comparatively narrow seas and a restricted zone of major operations made possible of accomplishment with some hundreds of thousands of mines (average cost, £400) what would in many cases and in many seas have been quite impracticable with as many millions of these difficult weapons.

****

The employment of submarines against submarines also forms a method of under-sea warfare which gives considerable scope for both daring and resource. It is of course quite impossible for one of these vessels when totally submerged to fight another in the same blind condition. But with just the small periscopic tube—or eye of the submarine—projecting above the surface, one craft can scout and watch for another to rise to the surface, thinking no enemy is near, in order to replenish her air supply for breathing or for recharging the electric storage batteries which supply the current for submerged propulsion.

When such a position obtains the submarine which comes unknowingly to the surface stands a grave danger of being torpedoed by her opponent. This actually occurred to at least one German U-boat during the Great War.

One or more submarines can also be employed around a slow-moving decoy ship. In this case they would have the advantage of being invisible until the actual moment of attack. The result of such a manoeuvre would be either a gun duel on the surface or the torpedoing of the attacking submarine by one or other vessel of the decoy's submerged escort.

It was a ruse of this kind which achieved success in the North Sea during the early stages of the war. A trawler was employed to tow a submarine by a submerged hawser. This mode of progress was adopted to enable the submarine to economise the strictly limited supply of electricity carried for under-water propulsion.

The trawler then cruised very slowly about, dragging the submarine under the surface behind her. In order to divert any suspicion which might have been aroused by her slow speed she was rigged so as to give the impression that a net was being towed, and the area of operations chosen was well-known fishing-ground.

In this curious way days were spent before the desired consummation was reached. Then a large U-boat came boldly to the surface and opened fire. Instantly the submarine astern of the trawler was released from the tow rope and forged ahead under her own electric engines. The commander of the surface decoy stopped his ship and commenced lowering the small life-boat carried. This was done in order to distract the attention of the Germans from the tiny periscope which was planing through the water to the attack.

A shell struck the trawler, carrying away her funnel, but did no other damage, and a few seconds later the water around the U-boat rose up in a vast upheaval of white. The plan had succeeded, and when the air cleared of the smoke from the trawler's damaged stack there was nothing afloat on the surface of the sea around—except an ever-widening patch of oil and bubbles.

A few minutes later the thin grey line of the British submarine rose above the swell some five hundred yards distant from the scene of her triumph.

Another means by which one subaqueous fleet can attack another is by laying mines in the seas around the enemy base.

****

These simple methods formed what may be termed the backbone of the widespread anti-submarine operations during the Great War, but with the experience gained and the brains of almost every nation focussed on the problem of providing an effective counterblast to the under-water warship, there can be little doubt that in the next great naval conflict new and more scientific means of attacking these pests of the sea will have been perfected, though what degree of success they will attain in the stern trial of war the future alone can tell.[7]

---

7. For a careful study of the effect of the submarine on the old theories of sea power see *Submarines and Sea Power*, by Charles Domville-Fife (Messrs George Bell & Sons, Ltd., London, and Messrs Lippincotts, New York.).

CHAPTER 11

The Mysteries of German
Mine-Laying Explained

To those unversed in modern war it may have appeared strange that, although the Allied navies held command of the sea from the opening of the Great War in 1914 to the signature of Peace in 1919, the Germans were nevertheless able to lay several thousand mines every year off the coasts of England, France and even the most distant colonies and dominions. It often occurred that harbour entrances and narrow fair-ways were repeatedly mined, notwithstanding a vigilant day-and-night watch from the bridges, look-outs and decks of many patrol ships cruising or listening in the vicinity.

The explanation is that the mines were laid by large submarines capable of approaching the coast, laying their deadly cargo from specially constructed stern tubes and retreating to comparative safety far out in the broad ocean, without rising more than momentarily to the surface for the purpose of observation.

This, it may be said, did not absolve the ships listening on their hydrophones, who should have been able to detect the approach of a submarine from the sound of her engines. During the first year of war the hydrophone was a very imperfect instrument, and although the sound might be heard it was quite impossible to tell from what direction it was coming. Later on, when the listening appliances had been greatly improved, there still remained two detrimental factors. The noise of breakers beating against rocks, sands or other obstructions destroyed much of the value of these instruments when used close inshore. On

A CAPTIVE MINE-LAYING SUBMARINE: U.C. 5 OFF TEMPLE PIER, LONDON

dark and rough nights the roar of wind and sea and the lurching of the vessel rendered subaqueous sounds extremely difficult to detect; and in a fair-way or channel used by surface shipping it was frequently impossible, even in fine but dark weather, to tell if the sound coming up from the sea emanated from a surface ship or a submarine.

If, in the latter case, the patrol ship started her own engines and moved forward in the darkness to ascertain from whence the noise came, she gave away her presence to the hostile submarine, *also fitted with listening appliances.* Whereas if she remained still and waited for the enemy to approach, mines might be laid in the meantime across important fair-ways which it was her duty to guard.

German mine-laying submarines were designated U–C boats, and often these vessels would employ a ruse in order to lay their mines in safety. Sometimes a decoy would draw the patrols away on a fruitless chase while the mines were being launched from the tubes of another U–C boat. In one case a big armed steamer was attacked with torpedoes while mines were being laid across the line of advance by which a flotilla of warships would be likely to come out to her aid from a near-by base.

In these and other ways over 3000 mines were laid off the British coast in one year. There were also several raids by surface mine-layers, which succeeded in eluding the network of patrols in the fogs and snows which prevail in the North Sea during several months out of every twelve. The two most important of these were the cruises of the *Wolfe* and the *Moewe*. The former vessel left Germany during the November fogs of 1916, and, by skirting the Norwegian coast, succeeded in passing the British patrol flotillas. She carried 500 mines, and after crossing the North Sea in high latitudes, proceeded down the mid-Atlantic until off the Cape of Good Hope, where the first mine-field was laid. She then crossed the Indian Ocean, laying fields off Bombay and Colombo.

It was in these seas that she succeeded in capturing a British merchantman. Placing a German crew and a cargo of mines aboard, she sent the prize to lay a field off Aden, while she herself

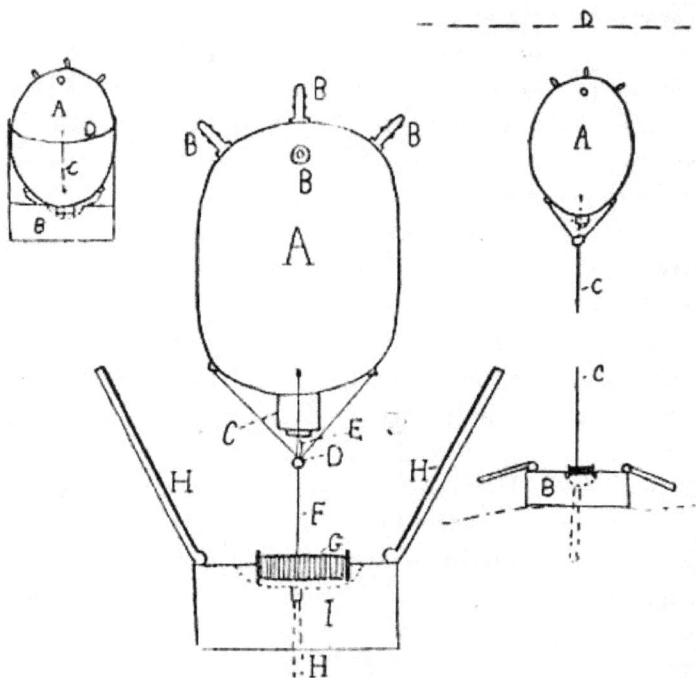

FIG. 23.

A typical German mine and sinker. *A*. The mine-casing containing about 300 lb. of high explosive, and the electric firing device which is put in force when the horns *B* are struck and bent by a passing ship. *B*. Horns, made of lead and easily bent if touched by a surface ship, but sufficiently rigid to resist blows by sea-water. *C*. Hydrostatic device, operated by the pressure of the water at a given depth, rendering the mine safe until submerged. *D*. Slings holding mine to mooring rope *F*. *F*. Mooring rope to reel in sinker. *G*. Reel of mooring wire, which unwinds when the mine floats to the surface. *H*. Iron supports held together (as in small left-hand diagram) by a band round the mine-casing. The mine goes overboard and sinks like this to the bottom. The band is then released by a special device, and the supports drop away, leaving the mine free to float to the surface (as in small right-hand diagram). *I*. A heavy iron sinker which acts as an anchor, holding the mine in one position.

proceeded to New Zealand. In these far-distant waters another field was laid, and a few months later the last of her cargo was discharged off Singapore. From this time onward she became a commerce raider.

The *Moewe* left Germany in December, 1916, and crossed the North Sea amid heavy snow squalls. Proceeding into the North Atlantic, she awaited a favourable opportunity to approach the British coast. This came one wild January night with a rising gale

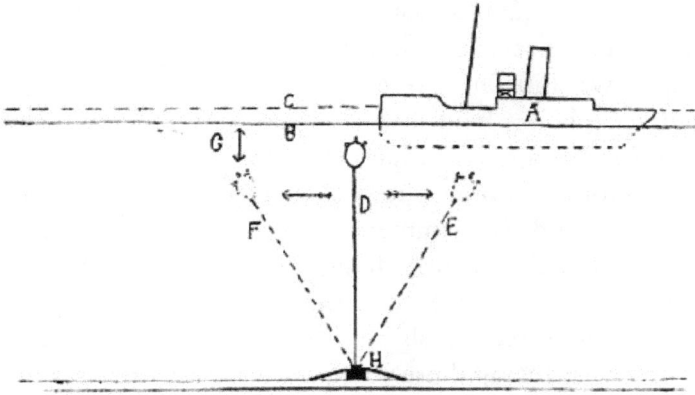

FIG. 24.

Diagram illustrating the effect of tide on a moored mine. A vessel is approaching a mine
D, moored to the bottom by a sinker *H*. The distance from the top of the horns of the
mine to the surface of the sea is approximately 5 feet at low tide, and as the vessel's
draught is 7 feet she would strike the mine. If, however, the same vessel passed over the
same mine a few hours later, at high tide, the level of the sea would have risen 5 feet,
and the mine would then be 10 feet below the surface; in which case the ship would
just pass over in safety. This is known as the "tide difficulty." There is, in addition, the
"dip" of the mine due to the strength of the tidal current. *E* and *F* show what is meant
by the dip of a mine. It is the deflection from the vertical caused by the ebb and flow
of the tide. It frequently causes a mine-field to be quite harmless to passing surface craft
except during the period of slack water between tides.

and a haze of snow. All her mines, about 400 in number, were
laid off the Scottish coast in the teeth of a nor'wester. Then, with
the "jolly Roger at the fore," she steamed out on to the wastes of
sea lying between the New World and the Old.

****

We now come to the mines themselves and the method of
laying them both above and below the surface.

A good idea of the shape, size and general characteristics of
these weapons will be obtained from the accompanying diagrams. On being discharged into the sea they automatically ad-
just themselves to float about ten feet below the surface (according to tide) and are anchored to the bottom by means of
a wire mooring rope attached to a heavy sinker. To describe
here the mechanical details of all the different types of German
submarine mines would occupy many pages with uninteresting
technical formulae. It is sufficient to say that they carried an

explosive charge (200 to 400 lb. of T.N.T.) sufficient to blow to pieces vessels of several hundred tons and to seriously damage the largest warship. They were intended to float a few feet below the surface—being held down by the mooring rope—but, as there was no means of compensating for the rise and fall of the tide, many of them often showed their horns above the surface at low water and were immersed too deep to be of much use against any but the deepest draught ships at high tide. A reference to Fig. 24 will make this difficulty clear.

There was scarcely a ship afloat in the zone of operations which did not, during those years of storm, sight one or more of these hateful weapons with their horns showing above the surface. Motor launches were employed to scout for them during the hour before and the hour after low water. In this way many hundreds were discovered and destroyed almost as soon as they had been laid. One badly laid mine, which shows on the surface when the tide ebbs, will often give away a whole field of these otherwise invisible weapons, and the work of sweeping them up and destroying them is then rendered comparatively easy.

The effect of strong tides on a moored mine is considerable, and will render a field quite harmless for several hours out of every twenty-four. The reason for this is best described with the aid of a diagram.

It will be seen from the above that the mine will not remain vertically above its sinker when there is a tide, but will incline at an angle determined by the strength of the current, which, if considerable, will press the weapon down much deeper than the keel of any ship (see Fig. 24). When the tide turns the mine will first regain its true perpendicular position and then incline in the opposite direction, accommodating itself to the ebb and flow. From this it will be apparent that in places where there is a strong current or tide a mine-field is only dangerous to passing ships of shallow or medium draft for a few hours (during slack water) out of the twenty-four. Between the ebb and the flow of a tide there is a short period when the water is almost still. Then the movement begins to set in from the opposite direction and gradually gains in speed until about one hour before high or low tide. This period of what is known as "slack water" varies considerably in different places and different weather conditions, but plays an important part in all minesweeping operations.

In this way many a ship has passed over a mine-field all unconscious of the fate which would have befallen her had she traversed the same area of sea an hour or so earlier or later.

Mines which break adrift, or are laid without moorings of any kind, are called *floating mines*. The latter are a direct violation of International Law, as they cannot be recovered when once they have been laid, and become a danger to neutral as well as to enemy shipping. The laws of civilised warfare also require even a moored mine to be fitted with some mechanical device which renders it safe when once it has broken adrift from the wire and heavy sinker which holds it in a stated position. The reason for this humanitarian rule is that neutrals can be warned not to approach a given area of sea in which there are moored mines, but if these weapons break adrift—as they frequently do in heavy weather—and float all over the oceans, they would seriously endanger the lives and property of neutral states unless something were done to render them innocuous.

The total disregard of all the laws and customs of civilised warfare by the Germans in 1914–1919 has now been so well established that it seems almost unnecessary to give yet another instance of this callousness. In the case about to be quoted, however, there is, as the reader will observe, an almost superlative cunning.

Any cursory examination of a German moored mine will show that there is a device fitted ostensibly to ensure the weapon becoming safe when it breaks adrift from its moorings and thus complying with The Hague Convention. For several months after the outbreak of war it puzzled many minesweeping officers and men why, with this device fitted, every German *floating* or *drifting* mine was dangerous. A few, relying on these weapons being safe when adrift, had endeavoured to salve one and had paid for the experiment with the lives of themselves and their comrades. This caused every mine, whether moored or adrift, to be regarded by seamen as dangerous, notwithstanding the oft-repeated assurances that German mines fulfilled all International requirements in this respect. Then a mine which had broken away from its moorings was successfully salved, in face of the great danger involved, and the truth came out.

A device *was* fitted to render it safe, but, with truly Hunnish ingenuity, the metal out of which an essential part of this appliance was made was quite unable to bear the strain imposed by its work, and, to make doubly sure, another part was half filed through. The result was that, instead of rendering the mine safe when torn from its moorings by rough seas, the essential parts broke and left the mine fully *alive*.

Any discovery such as this—*only made at the great risk of salving a live mine*—could be easily explained away by German diplomacy as faulty workmanship in a particular weapon, reliance being placed on the fact that not many mines could be salved in this way without heavy loss of life; but numbers were recovered in spite of the dangers and extraordinary difficulties of such operations, and the guilt was for ever established in the minds of those who sail the seas.

Little need be said here regarding the method of laying mines from surface ships like the *Wolfe* and *Moewe*. The weapons were arranged to run along the decks on railway lines and roll off the stern, or through a large port-hole, into the sea as the vessel steamed along.

With submarine mine-layers or U-C boats the method was, however, much more complicated and needs full description. Each vessel was fitted with large expulsion tubes in the stern and carried some eighteen to twenty mines. These weapons, although similar in their internal mechanism to the ordinary mine, were specially designed for expulsion from submerged tubes or chambers.

The mines were stored in the stern compartment of the submarine, between guide-rails fitted with rollers. They were in two rows and moved easily on the well-greased wheels. The loading was accomplished through water-tight hatchways in the deck above. In order to expel these mines from the interior of the submarine when travelling under the surface each weapon had to be moved into a short expulsion tube or chamber, the inner cap of which was closed when a mine was inside, and the outer or sea-cap opened. A supply of compressed air was then admitted into the back of the tube and the mine forced out into the open sea, in the same way as a torpedo is now expelled from a submerged tube.

Before another mine could be launched the sea-cap had to be closed, the water blown from the tube, the inner cap opened and a second mine placed ready in the chamber. This, however, did not end the difficulty of laying mines from submarines. The increase in the buoyancy of the boat, due to the loss of weight as each mine was discharged into the sea, had to be instantly and automatically compensated by the admission of quantities of sea-water of equal weight into special tanks, hitherto empty, situated below the mine-tubes. If this had been neglected the submarine would have come quickly to the surface, stern uppermost, owing to the lightening of the hull by the expulsion therefrom of some fifteen weapons weighing many hundreds of pounds each.

When the mine was clear of the submarine it sank to the bottom, owing to the weight of the sinker or anchor. After a short immersion, however, a special device enabled the top half, containing the charge of explosive and the contact firing horns, to part company with the heavy lower half, composed of the iron sinker and the reel of mooring wire. The explosive section then floated up towards the surface, unwinding the wire from the sinker.

Each mine being set, before discharge, to a certain prearranged depth (obtained by the captain of the U-C boat either by sounding wires or from special charts showing the depth of water in feet), the weapon could not rise quite up to the surface, being checked in its ascent, when ten feet from the top, by the mooring wire refusing to unwind farther.

This may sound a little involved, but a careful study of the accompanying diagrams will make the various movements of the mine and its sinker, after leaving the submarine, quite clear to the lay reader.

There were also other types of mines employed. Some were

fitted with an automatic device which was actuated by the pressure of the water at a set depth. These weapons could be expelled from submarines without the necessity of knowing and adjusting the depth at which they were to float below the surface. A mine of this pattern rose up, after discharge from the tube, until the pressure of water on its casing was reduced to 4 ½ lb. per square inch (the pressure which obtains at a depth of ten feet below the surface[8]), and there the weapon stopped, waiting patiently for its prey.

Another kind of mine was of the floating variety—tabooed by The Hague Convention—which drifted along under the surface with no moorings to hold it in one position.

Now that the reader is familiar with the mines themselves and the actual methods of laying them, we can pass on to a brief review of the German mine-laying policy during the Great War.

The submarine offensive reached its maximum intensity in 1916-1917, during which period no less than 7000 mines were destroyed by the British navy alone.[9] Of this number about 2000 were drifting when discovered. There was, with one small exception, no portion of the coast of the United Kingdom which was not mined at least once during those eventful *two* years, the unmined area being undoubtedly left clear to facilitate a raid or invasion. About 200 minesweeping vessels were blown up or seriously damaged, but the losses among the Mercantile Marine were kept down to less than 300 ships out of the 5000 sailings which, on an average, took place weekly.

The heavy losses inflicted on the enemy's submarine fleets in 1917 marked the turning of the tide, and from that date onwards there was a steady but sure reduction in the number of mines laid.

During the first twelve months of the intensified submarine war the Germans concentrated their mine-laying on the food routes from the United States, the sea communications of the

---

8. The question of water pressures and many other problems of submarine engineering relating to under-water fighting are fully treated in *Submarine Engineering of To-day*, by the Author.

9. A few of the 7000 were British mines no longer required in the positions in which they had been laid.

Grand Fleet off the east coast of Scotland and the line of supply to France. Then, when they commenced to realise the impossibility of starving the sea-girt island, and the weight of the ever-increasing British armies began to tell in the land war, the submarine policy changed to conform with the general strategy of the High Command, and the troop convoy bases and routes were the objects of special attack.

The arrival in Europe of the advance guard of the United States army caused another change in the submarine strategy. From that time onwards the Atlantic routes assumed a fresh importance and became the major zone of operations.

In the first year of the war the U-C boats discharged their cargoes of mines as soon as they could reach their respective areas of operation. The mines were usually laid close together in one field, frequently situated off some prominent headland, or at a point where trade routes converged. Then the enemy learned to respect the British minesweeping and patrol organisation, and endeavoured to lay their "sea-gulls' eggs" in waters which had been recently swept, or where sweeping forces appeared to be weak in numbers.

When this failed they played their last card, scattering the mines in twos and threes over wide areas of sea. To meet this new mode of attack large numbers of shallow-draught M.L.'s were employed to scout for the mines at low water.

It was about this time that the great Allied mine barriers across the entrances and exits to and from the North Sea were completed and the losses among the U and U-C boats became heavy. A rapid abatement in the submarine offensive soon became apparent, and utter failure was only a matter of time.

CHAPTER 12

The Mysteries of Minesweeping Explained

The task which confronted the naval minesweeping organisations in the years succeeding 4th August 1914 was an appalling one. Any square yard of sea around the 1500 miles of coast-line of the British Isles might be mined at any moment of any day or night. There were, in addition, the widely scattered fields laid by surface raiders like the *Wolfe* and the *Moewe*, which, as described in a previous chapter, extended their operations to the uttermost ends of the earth. A wonderfully efficient patrol of the danger zones had its effect in reducing the number of submarine mine-layers available to the enemy and in rendering both difficult and hazardous the successful execution of their work, but neither a predominant and subsequently victorious fleet, nor an equally skilful and alert patrol, could guarantee the immunity of any considerable area of sea from mines.

The Germans laid many thousands of these deadly and invisible weapons in the 140,000 square miles of sea around the British Isles *alone* in the face of over 2000 warships. To search for these patches of death in the wastes of water may well be likened to exploring for the proverbial "needle in a haystack." Yet the sweepers, whose sole duty it was to fill this breach in the gigantic system of Allied naval defence, explored daily and almost hourly, for over four years, the vast ocean depths, discovering and destroying some 7000 German mines, with a loss of 200 vessels of their number. The result of this silent victory over one of the greatest perils that ever threatened the Sea Empire was that some 5000 food, munitions and troop ships were able to enter and leave the ports of the United Kingdom *weekly* with a remarkably small percentage of loss from a peril which might easily have proved disastrous to the entire Allied cause.

This, then, in broad outline, was the task which confronted this section of the naval service, and its successful accomplishment forged a big link in the steel chain encompassing the glorious victory.

Before passing on to describe the ships and the appliances used it is first necessary to give a more detailed account of the operations generally included under the heading of minesweeping. As it was impossible to tell exactly where mines would be laid from day to day, an immense area of sea had to be covered by what was known as *exploratory sweeping*. This consisted of many units of ships emerging from the different anti-submarine bases almost every day throughout the year and proceeding to allotted areas of water, where they commenced sweeping north, south, east or west, in an endeavour to discover if the areas in question were safe for mercantile traffic. If no mines were discovered that particular area would be reported safe, but if only one of these weapons was cut from its mooring by a sweep-wire the area would be closed to merchant ships until the sea around was definitely cleared of the hidden danger. This system of open and closed areas entailed an enormous amount of efficient administrative staff work apart from the actual sweeping, and its success was partly dependent upon the vigilance of the patrols employed to divert shipping from dangerous patches of sea.

When a mine-field was discovered which interfered with the free movement of a large number of ships a big concentration of sweepers from all the adjacent bases was ordered by telegraph and wireless. The area was isolated by patrols and the mines swept up. In one field no less than 300-400 mines were known to have been laid. Finally a further exploratory sweep was made, and if nothing further was discovered the area was again opened to traffic, and the sweepers turned their attention either to routine duties or to the clearance of another field.

The entrance to every important harbour was swept once

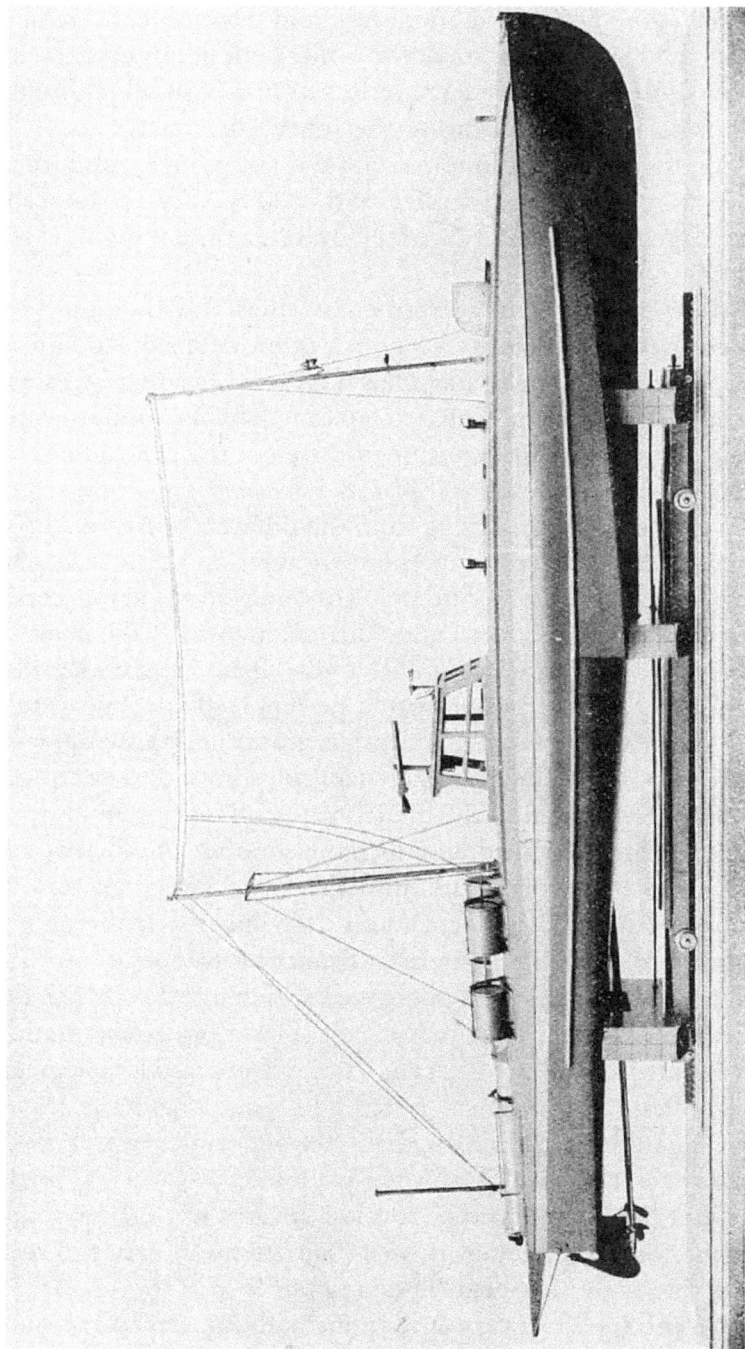

MODEL OF A COASTAL MOTOR BOAT (55 FT.) WITH TORPEDO AND FOUR DEPTH CHARGES

or twice a day, and all convoys had sweepers ahead of them when they left or entered such confined waters. The seas adjacent to harbours and naval bases were searched at low water for mines which might be showing above the surface. Around the anchorage of the Grand Fleet in Scapa Flow a wide belt of sea was kept clear of mines so that at any moment the fleet could reach blue water without risk from these weapons. The same precautions were taken off the Firth of Forth for the benefit of the battle cruisers, and outside Harwich for Admiral Tyrwhitt's light forces.

A passage known as the "war channel"—about which more will be said later—extending from the Downs to Newcastle, was swept daily by relays of sweepers operating from the anti-submarine bases along this 320 miles of coast-line. This buoyed and guarded channel formed a line of supply for the great fleets in the north.

Each big fighting formation was provided with a special flotilla of fast fleet sweepers, which were capable of clearing the seas ahead of the battleships and cruisers moving at 20 knots. This was a separate organisation to what may be described as the routine sweeping of the trade routes. These vessels were always within call of the fleets they served.

It has been estimated that over 1000 square miles of sea were swept daily by the anti-mine fleets of the British navy during the four years of war. This may not sound a very stupendous figure compared with the area of the danger zone, but in practice it necessitated terribly hard work from dawn to dusk by several thousand ships and many thousands of men in summer heat and winter snow.

There was in addition to all this the clearing of British minefields no longer required in the positions in which they had been originally laid. This was not entirely an after-the-war problem, for although the great mine barriers were left until peace was assured, there were fields of minor importance which had to be cleared to meet new situations as the years of war passed swiftly by. A notable instance of this was the destruction of a big field of some 400 mines off the Moray Firth.

The foregoing refers only to the minesweeping in the principal danger zones in British waters, no account being taken of the work carried out by Allied vessels in the Mediterranean, off the coasts of France, Italy, Greece, Gallipoli, and in such distant seas as those washing the shores of New Zealand, Australia, Hong-Kong, Japan, Singapore, Bombay, Aden, the Cape of Good Hope, the United States, Eastern Canada, West Africa and Arctic Russia, in all of which mines were laid by surface raiders like the *Wolfe*, and afterwards located and cleared by Allied warships.

From the foregoing some idea of the gigantic nature of the task will be obtained, and we can pass on to a more detailed account of the actual work. Minesweeping may be divided into eight well-defined sections, as follows:

(1) *Fleet Sweeping.*—Keeping clear the sea routes of the battle fleet.

(2) *Exploratory Sweeping.*—Searching the sea for isolated groups or fields.

(3) *Routine Sweeping.*—The daily or weekly sweeping of areas, channels and coastal trade routes, largely used by shipping.

(4) *Clearing Large Mine-fields.*—Big concentrations of ships to rapidly clear important routes temporarily blocked by large mine-fields.

(5) *Special Shallow-Water Sweeping.*—Such as that carried out off the Belgian coast by specially constructed shallow-draught ships, frequently with single-ship sweeps.

(6) *Convoy Sweeping.*—Precautionary sweeping in front of incoming and outgoing convoys. This was regularly done even if the fair-way was covered by routine sweeping.

(7) *Harbour Sweeping.*—Precautionary sweeping usually carried out by small craft at big naval bases such as Portsmouth (Spithead) and Rosyth (Firth of Forth) inside the submerged defences.

(8) *Searching at Low Tide.*—This was done by shallow-draught vessels of the M.L. type in order to locate badly laid mines which might project above the surface at low water. Several hundred were discovered in this way.

In order to carry out these duties efficiently the heterogeneous fleet of minesweepers was divided into small fleets stationed at the numerous anti-submarine bases, and these were again subdivided into units of ships especially adapted for the different classes of work. Each *pair* of vessels had to be more or less alike in size, draught, speed and manoeuvring ability to enable them to work efficiently in dual harness. Consequently there were complete units of vessels specially constructed for dealing rapidly with discovered mine-fields and for use with the battle fleets. Shallow-draught vessels of the motor launch type for work in the shallow water off the Belgian coast. Converted pleasure steamers of the usual Thames, Mersey and Clyde type for convoy sweeping. Motor launches for clearing fair-ways and for searching at low water. Flotillas of trawlers and drifters for the hard and monotonous routine sweeping on the important coastal trade routes. They comprised in all several thousand ships engaged solely on this work.

At each important base there was a Port Minesweeping Officer (P.M.S.O.), with one or more assistants, whose duty it was to administer, under the command of the S.N.O., the fleets in the attached area, and to furnish preliminary telegraphic and detailed reports to the Minesweeping Staff at the Admiralty, who issued a confidential bi-monthly publication to all commanding officers which was a veritable encyclopaedia of valuable information regarding current operations, events and enemy tactics. Attached to this department was a section of the Naval School of Submarine Mining, Portsmouth, where all knotty problems were unravelled and appliances devised to meet all kinds of emergencies.

Each unit of ships was under the command of a senior officer, responsible for the operations of these vessels, and where big fleets were engaged a special minesweeping officer was placed in supreme command. Only by close co-ordination of effort from the staff at Whitehall and elsewhere to the units at sea could this gigantic work have been expeditiously accomplished. It frequently happened that any delay due to very severe weather in clearing a field or area meant complete stoppage or chaotic dislocation of the almost continuous stream of merchant shipping

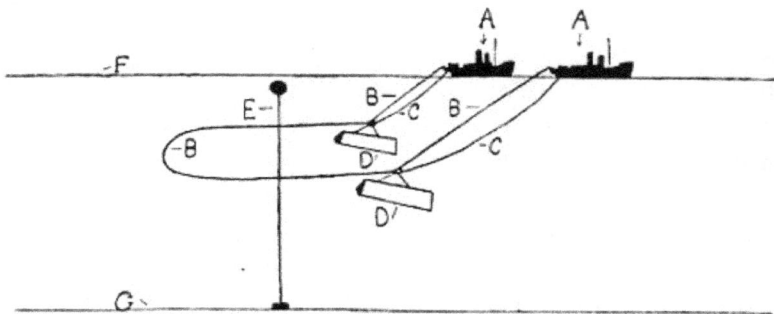

FIG. 25.

Diagram showing the form of apparatus principally used by British minesweepers. *AA*. Sweeping vessels. *BB*. Sweep-wire. *CC*. Wires holding kites. *DD*. Kites which hold sweep-wire at correct depth below the surface by their "kite-like" action when being towed through the water. *E*. Mine and mooring. *F*. Surface of the sea. *G*. Sea-bed.

entering and leaving a big harbour, which, in turn, disorganised the adjacent harbours to which ships had often to be diverted. It disturbed the railway facilities for the rapid transport of the food or raw materials from the coast to the manufacturing centres, from the sugar on the breakfast-table to the shells for the batteries in France. One hour's delay in unloading a ship may mean three hours' additional delay on the railways, the loss of a shift at a munitions works and a day's delay in a great offensive. It is a curious anomaly, made vividly apparent to those in administrative command during the past years of stress, that the more perfect the organisation the greater the delay in the event of a breakdown in the system.

There were various methods of minesweeping, but in all of them the object was to cut the mooring wire of any mine that came within the area of the sweep and so cause the mine itself to bob up to the surface, where it could be seen and destroyed by gun-fire. In order to encompass this many kinds of minesweeping gear were devised and given practical trial during the war. The one most generally used, however, was the original but vastly improved sweep. This consisted of a special wire extended between two ships and held submerged by a device known as a kite. This apparatus is best described diagrammatically (Fig. 25).

There was, however, another type of sweep used for exploratory work and also for sweeping in shallow water. It was a one-

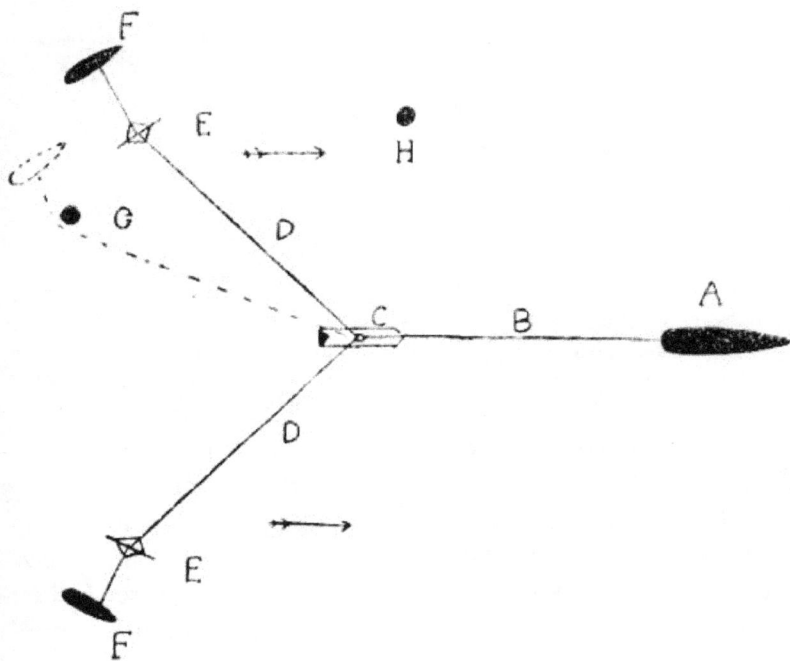

FIG. 26.

Diagrammatic sketch showing principal parts of a single-ship sweep. *A.* Towing vessel.
B. Tail wire. *C.* Kite holding sweep-wires *D* at correct depth below the surface. *D.* Light
sweep-wires held at an angle by spars *E* and surface hydroplane floats *F.* The dotted lines
show how either arm of the sweep swing towards the centre line when exposed to the
pull of a mine. This movement of the hydroplane floats indicates to those on board the
sweeping vessel that a mine has been caught. The mine *H* slides down the sweep-wire
until the mooring is cut at *G,* and the mine floats freely to the surface.

ship sweep (*i.e.* required only one vessel to drag it), and this can
also be best described by a diagram (Fig. 26).

It will be observed that in all cases the object is to drag a
submerged wire through the water at an angle from the ship's
course until it encounters the mooring wire of a mine. When
this takes place it is the purpose of the sweep-wire to cut the
mooring wire and allow the buoyant mine to float up to the surface free of its sinker (see Fig. 27). In order to effect this various
kinds of hard wire with a cutting capacity were used as sweep-
wires, and also numerous mechanical devices, all of which are
more or less of a secret character; but the object remained the
same—to find and cut the mooring wire.

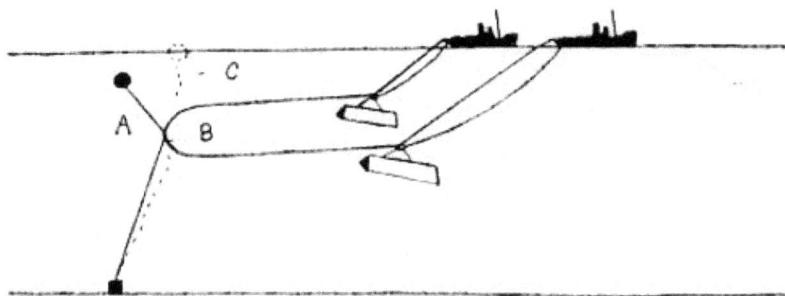

FIG. 27.

Diagram showing mine mooring being cut by sweep-wire. *A*. Mine-mooring wire. *B*. Hard and cutting face of sweep-wire. The dotted lines *C* show how the mine floats to the surface by its own buoyancy when the mooring wire holding it down has been cut.

The introduction of what became known as "delayed action mines"—weapons held down on the sea-bed, after being launched, for varying periods of time, so that sweeping operations might take place above them without their being discovered; then, when the time for which the delay was set had expired, they rose to within ten feet of the surface and became a great danger to shipping in places recently swept and reported clear—caused a new form of sweep to be devised and used in waters where these mines were likely to be sown.

This type of sweep was known as a "bottom sweep," and generally consisted of a chain fitted into the bight of a sweep-wire and dragged along the sea-bed, the idea being to overturn the delayed mine and so upset its mechanism that it would either rise immediately to the surface or else remain for ever harmless at the bottom of the sea. In many cases the heavy chain passing over the horns of the mine would bend and make them useless, so destroying the efficiency of the mine even if it did eventually rise to the correct firing depth.

Into almost every operation carried out on or under the sea there enters the tide difficulty, and in all mining and mine-sweeping operations it is one of the most important factors to be considered. The effect of the tide on mine-laying has been dealt with in a previous chapter, and the same difficulties in reverse order are experienced when sweeping the sea for these invisible and dangerous weapons. It has already been shown that a vessel may sometimes pass safely over a mine at high water which would touch her sides or keel and explode if she passed over it at low water when the mine was nearer to the surface. All mine-sweeping vessels, therefore, need to be of comparatively shallow draught in order to reduce the risk of touching mines, but against this is the fact that shallow-draught ships, even if powerfully engined, have but little grip on the water and experience an undue loss of speed when towing a heavy sweep-wire. Such vessels can seldom operate in even moderately heavy weather owing to their rolling and pitching propensities. Therefore a vessel of medium—bordering on shallow—draught, with a fairly broad beam, is the best type. Here, again, is a difficulty. Minesweeping is a type of defensive warfare requiring a vast number of ships successfully to carry on against an enemy well provided with surface and submarine mine-layers, and not even the greatest naval power in the world could seriously contemplate maintaining a peace fleet of, say, 2000 such vessels in constant readiness. Therefore recourse has to be made, when war comes, to mercantile craft, which seldom possess all the desired qualities.

This is what actually occurred in every maritime country at war during the years succeeding August, 1914, and in order to meet the danger attending the use of passenger ships, trawlers and drifters, often with a considerable draught, minesweeping operations were, whenever possible, confined to the three hours before and the three hours after high water. Shallow-draught M.L.'s carried out the scouting for mines at low tide. It is difficult to see what would be the fate of a nation hemmed in by mines and devoid of a mercantile fleet sufficiently numerous to provide powerful sweeping units. The trawlers and pleasure steamers were a godsend to England in those years of intensive submarine warfare. This undeniable fact incidentally provides another example—if such is now needed—of naval power resting not entirely on fleets and dockyards, but on every branch and twig of maritime activity.

It is difficult to describe in small compass and non-technical language the various tactical formations employed in minesweeping operations. They were many and various. The Ger-

131

mans used their vessels in long lines, the ships being connected
together by a light wire-sweep plentifully supplied with cutting
devices, into which the mooring wire of the mine was expected
to obligingly slip. This method suffered from the serious drawback that if any part of the sweep-wire caught on a submerged
obstacle, such as a projection of rock, the whole line of ships
became disorganised. There were also many other objections to
this system, some of which will doubtless be apparent to the
thoughtful reader.

The formation usually adopted by British minesweepers was
that shown in Fig. 28, in which it will be observed that each
pair of ships is actually independent of the others, but is acting
in company with them, and that the pathway swept by one pair
is slightly overlapped by the following pair. In the event of an
accident to one ship the next astern can immediately let go its
own end of sweep-wire and go to the rescue of any survivors. It
may be apropos to say here that the smaller class of minesweeper
is usually blown to pieces if she touches a mine.

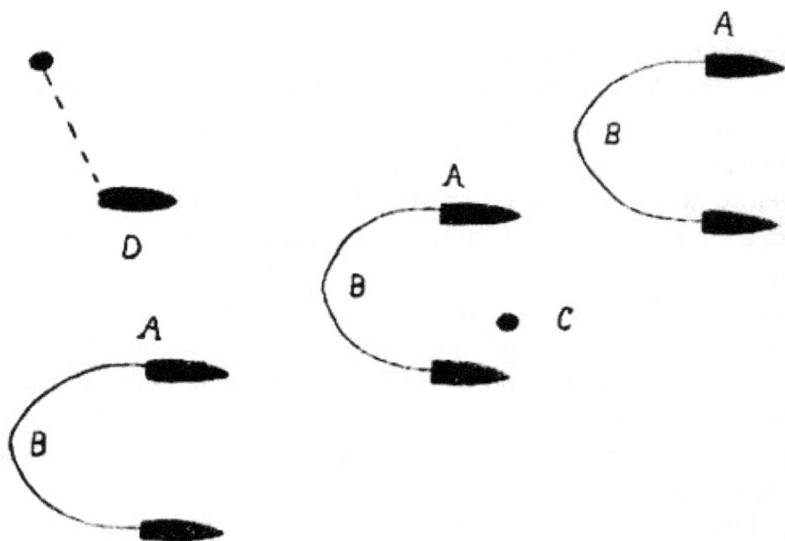

FIG. 28.

Plan showing the usual formation adopted by British minesweeping vessels. A. Three
pairs of sweepers. B. Sweep-wires. C. A mine entering the sweep of the second pair. D. A
vessel following the sweepers for the purpose of sinking by gun-fire the mines cut up.

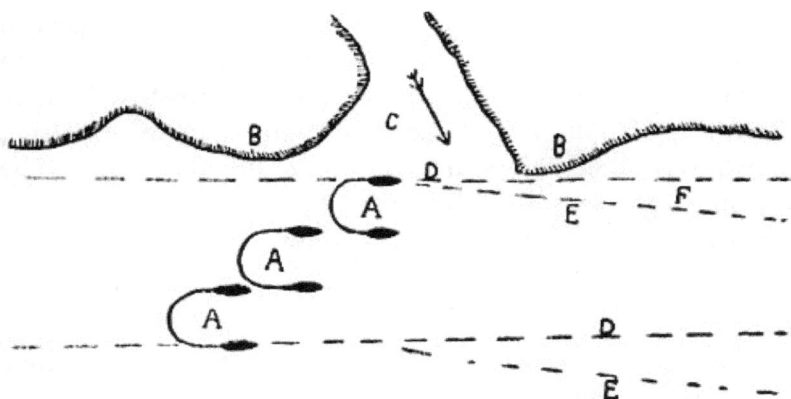

FIG. 29.

Diagram illustrating the effect of tide on minesweeping operations. *A*. The vessels sweeping along the coast-line *B*. A fast ebb-tide is coming down the estuary *C*. Unless an allowance was made for this tide and mark-buoys or ships were placed along the dotted course *D*, the sweepers would unknowingly drift seawards along course *E*, leaving a space *F* unswept and possibly dangerous to ships entering and leaving the estuary *C*.

The set of the tide is another important factor which has to be taken into serious consideration when plotting a sweep. This complication enters into every operation, and its salient points will be made quite clear by referring to Fig. 29.

The actual speed at which minesweeping operations are carried out depends greatly upon the engine-power of the sweepers themselves. In the case of trawlers and drifters it is seldom possible to drag the 300-600 feet of heavy wire through the water at a greater rate than 4 to 6 knots. M.L.'s can accomplish 8 knots with a lighter wire, while big fleet sweepers with engines of several thousand horse-power can clear the seas at 18-23 knots.

Sufficient has now been said to enable the reader to realise fully the arduous, exciting and often very hazardous nature of the work. Veteran sweepers listen for the loud hum of the wire which proclaims that a mine has been caught. Then comes an interval of a few seconds of suspense. Sometimes the mine bobs up within a few feet of the ship; at other times it is in the middle or bight of the wire, far astern, and half-way between the two sweeping vessels. When a mine is cut up a few shots from a 3-pounder, a shattering roar and the mine is destroyed. All that remains is a column of smoke reaching from sea to sky.

It frequently happened that the mine became entangled in the sweeping gear and was unknowingly hauled on board with the sweep. When this occurred the position was fraught with extreme peril. Any roll of the ship might cause an explosion which would shatter to fragments everything and everyone within range. Safety lay in lowering the sweep gently back into the sea—an extremely difficult operation on a rough day.

THE WAR CHANNEL

This carefully guarded fair-way consisted of a 320-mile stretch of sea, extending along the east coast of England from the Downs to Newcastle, which was marked on the seaward side by a continuous line of gigantic buoys, two miles apart. It was patrolled day and night by hundreds of small warships, and swept from end to end by relays of sweepers acting in conjunction with each other from the different anti-submarine bases along the coast.

The war channel formed a comparatively safe highway for all coastal shipping passing north or south through the danger zone, and vessels from Holland, Denmark, Norway and Sweden were able to cross the North Sea at any point under escort and proceed independently and safely along the British coast to whichever port could most conveniently accommodate them at the time of their arrival. It also relieved the terrible congestion on the railway lines between the north and south of England by enabling a coast-wise traffic to be carried on between the ports of London, Grimsby, Hull and Newcastle, as well as enabling the numerous Iceland fishing fleet to pass up and down the coast in comparative safety on their frequent voyages to and from the fishing grounds of the far north. From the naval or strategic point of view it more or less secured a line of supply for the Grand Fleet assembled in the misty north. Colliers, oilers, ammunition and food ships were able to proceed through the comparatively narrow section of the danger zone with a minimum of risk; and, had it been required, there was available a cleared passage for any squadron from the big fighting formations to come south at high speed to checkmate a bombardment or attempted landing on anything like a grand scale.

It may perhaps be wondered why *this* channel was not extended up the east coast of Scotland as far as Scapa Flow. In the first place, the North Sea widens considerably as the higher latitudes are approached, the coast of Scotland does not lend itself to a clearly defined channel and the heavy weather which prevails for so many months in the year made the maintenance of gigantic buoys and their moorings almost impossible. Secondly, there were various systems of mine defences in this area, and, although not defined by a chain of buoys, the passage north from Newcastle to the Scottish islands was, in actual fact, maintained by a vast organisation of patrols and sweepers, but over this section of sea supply ships for the Grand Fleet were nearly always under escort. The area from the Scotch to the German coast was looked upon more as a possible battleground for the fleets at war than as a route for merchant shipping, owing to the comparatively few big commercial harbours along the eastern shore.

Laying the moorings of over 150 gigantic buoys in fairly deep water, exceptionally prone to sudden and violent storms, was in itself a noteworthy feat of submarine engineering. The chains and anchors had to be of great strength, and the whole work, which occupied many weeks, was carried out in waters infested with submarines and mines.

The task of sweeping this vast stretch of sea almost continuously for four years was by no means either straightforward or without risk. The Germans, when they discovered the existence and purpose of this channel, sought to turn it to their own advantage by systematically laying mines around the moorings of the mark-buoys, where they could only be swept up with great difficulty, owing to the sweep-wires fouling the moorings of the buoys. This stratagem had to be answered by the creation of "switch lines," or small sections of false channel marked by buoys, while the real channel was only outlined on secret charts. In this way the preservation of the war channel and its use for misleading and entrapping U and U-C boats became a semi-independent campaign, in the same way as that which surrounded the great mine barrages and other activities of the anti-submarine service.

MINE PROTECTION DEVICES

It is an axiom of war that new weapons of attack are invariably met by new methods of defence. The mine was no exception to this rule, although up to the present time the various antidotes are in all cases only partial remedies. During the years of war, with the brains of a maritime nation focused on the subject, there were naturally many devices suggested and tried for protecting ships from mines. The great majority of these suggestions may be classified in two groups: (1) Those which sought to deflect the mine from the pathway of the ship; and (2) those which sought to minimise the result of the explosion. One method from each of these groups was adopted with various modifications to suit different classes of ships.

In the first group came the *Paravane*, which had as its basis the suspension of a submerged wire around the bow of a ship, which caught and deflected the mine-mooring wire before the horns of the mine itself could reach the sides of the ship. It also cut the mooring and enabled the mine to rise to the surface and be destroyed by gun-fire.

In order to understand this appliance it is first necessary to know what is the action of the majority of moored mines on coming in contact with a ship. It seldom happens that a vessel strikes a mine dead on the bow or stem-post. The cushion and dislocation of water formed by a big and fast ship around its bows is usually sufficient to cause the mine to swing a few inches away from the bow and to return and strike the ship several feet back on the port or starboard side. A careful study of Fig. 30 will show how this is prevented by the deflecting wires of the paravane.

The paravanes themselves are submerged torpedo-shaped bodies which hold the wires under the surface and away from the ship's side, deriving their ability to do this from the speed at which they are being towed, submerged, by the ship itself. A

A PARAVAN: HOISTING IN THE STARBOARD PARAVANE OF
THE P.V. MINE-DEFENCE GEAR.

piece of string through the axle hole of a small wheel, which is
then placed on the ground and pulled along, will give a good
idea of the action of the paravane against the passing water.

It is not possible to give here the exact details of this highly
ingenious device upon which so much scientific and practical
attention was wisely bestowed, but sufficient has been said to
enable the reader to form a clear conception of how the mine
was caught and held away from the ship's side by the deflecting
wire of the paravane.

This device, in one of its many forms, was fitted not only
to warships, but also to many hundreds of merchantmen, and
was known to have saved thousands of tons of valuable shipping and cargo.

Among those devices which had for their object the mini-
mising of the result of a mine explosion may be mentioned the
"Blister System" so successfully employed in the construction
of monitors and other big ships, the idea being to surround the
inner hull with an outer casing which received the effect of the
explosion of either a mine or torpedo and left the inner or real
hull of the ship water-tight. Its one weak feature was that it reduced the speed of the ship and the ease with which she could

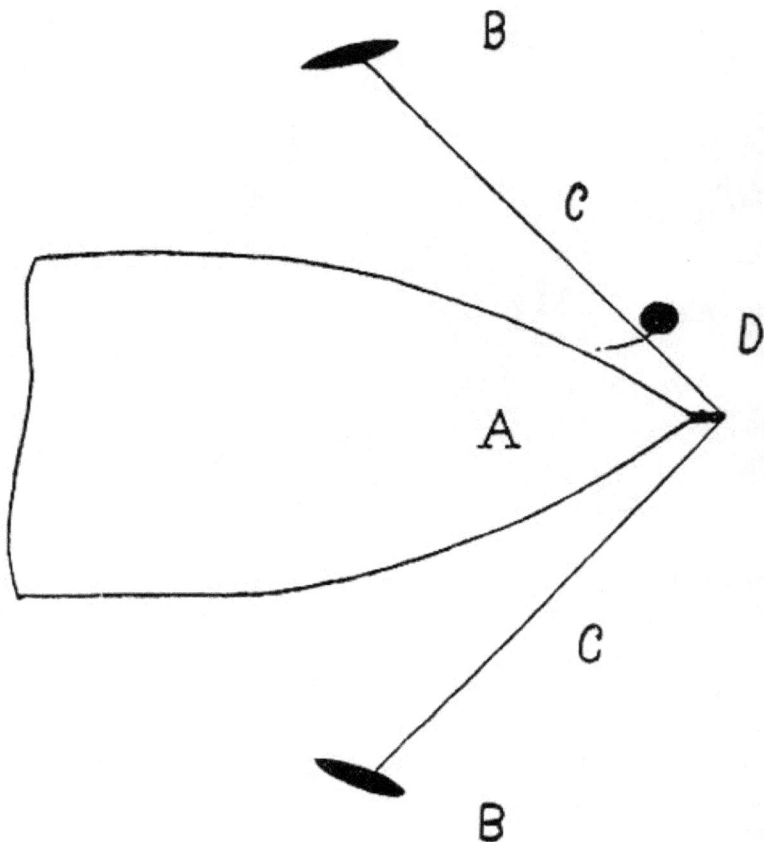

Fig. 30.

Plan showing the chief characteristics of the paravane mine defence gear. *A*. The bow of the ship. *B*. The paravanes being towed submerged at an outward angle. These appliances maintain a fixed depth below the surface and hold the ends of the deflecting wires *C* well away from the ship's sides. *C*. The submerged deflecting wires, held at one end by a short projection from the ship's stem-post below the water-line, and at the outer end by the submerged paravanes. *D*. A mine and its mooring caught by the deflecting wires and held away from the ship. In such a case it would slide down the deflecting wire towards the paravane, where the mooring would be cut and the mine would float to the surface.

be manoeuvred. In future types of large and heavily armed ships this drawback will undoubtedly be largely overcome by an increase in engine-power made possible by the development of engineering science.

The "blister," although outwardly forming a continuous

structure round the entire vessel, extending well above and below the water-line, tapered off towards the bows and stern, and was subdivided into different compartments. In this way an explosion against one section did not necessarily damage any other part. The British monitors which so successfully bombarded the Belgian coast and the fortifications of the Dardanelles were fitted with blisters, and more than one of them owed their salvation to this means.

Chapter 13

The Mine Barrage

What undoubtedly forms the most effective counter to unrestricted submarine warfare is the explosive mine barrage, as employed against the German U-boats in the North Sea and the Straits of Dover.

The practicability of these barrage systems depends, however, very largely upon the following factors:—(1) the geographical features of the area of operations; (2) the hydrographical peculiarities of the seas in which the mines have to be laid; (3) the number of properly equipped mine-laying vessels available; (4) an adequate and highly trained personnel; and (5) the mechanical skill and manufacturing power of the nation employing the system.

There are several forms of mine barrage. One is simply an elongated mine-field laid across a narrow sea to prevent the safe passage of hostile surface craft. In this case the mines are laid in the ordinary manner and at the ordinary depth below the surface. The anti-submarine barrage, however, consists of an enormous number of mines, laid *at a considerable depth below the surface* and in such formation as to ensure that a submarine attempting to pass through the cordon *while submerged* would inevitably collide with one or more of them.

With this latter form of barrage the surface of the sea is quite clear of mines and is comparatively safe for the unrestricted movement of a numerous patrol flotilla, which forms part of the system, the under-seas alone being made dangerous by the mines.

It will be apparent that if a hostile submarine base is enclosed by one or more of these barrages the under-water craft entering and leaving that base have the choice of travelling *submerged* across the danger zone and thereby risking contact with the mines, or of performing the passage *on the surface* and encountering the patrolling ships. In either case, the result is more likely than not to be the destruction of the submarine.

In most cases the exact position of the barrage would be unknown to the hostile submarines, which, even if running on the surface, would dive immediately on the approach of a patrol ship. The few lucky ones succeeding in getting safely through the cordon of deep-laid mines, or passing unnoticed the patrol of surface ships on their outward journey—as might be the case in fog—would have the same peril to face on the return to their base, and probably without the aid of thick weather. This double risk would probably have to be taken by every submarine in the active flotilla at least once a month, this being approximately the period they can remain at sea without replenishing supplies of fuel, torpedoes and food.

The object of the flotillas of shallow-draught patrol vessels operating in the vicinity of the deep mine barrier is twofold. Primarily their duty is to prevent the hostile submarines from running the blockade on the surface and, secondly, to prevent enemy surface craft from emerging from the base and sweeping clear a passage through the mine-field, or of laying counter-mines, which, on being exploded, would detonate some of the blockading deep-laid mines and so destroy a section of the barrier.

From this it will be apparent that a force of hostile submarines hemmed in in this way would run a double risk of losing a number of vessels on every occasion on which a sortie was made. This is what actually occurred to the German under-water flotillas in the years 1917–1919, and, in combination with the other methods employed by the Allied navies, was mainly responsible for the failure of the great under-sea offensive.

The only bases of the German navy being situated on the North Sea littoral, it was possible for the Allies to lay a vast mine barrier, stretching from the coast of Norway to the Scottish islands, and another smaller one across the Straits of Dover; also to concentrate in the vicinity of these two submarine "trench systems" a very numerous surface patrolling force, thus enclosing the thousands of square miles of sea forming what was sometimes boastfully referred to as the "German Ocean" in an almost impenetrable ring of steel and T.N.T.

Here let us consider the gigantic nature of the task that was successfully accomplished. The distance from the Norwegian coast to the Orkney Islands is approximately 600 miles. It was over this vast expanse of sea, bent at the eastern end so as to rest on the Heligoland Bight, that the system known as the "Northern Barrages" extended. No exact statistics of the actual number of mines used is at present available, but reckoning at the low rate of one mine to every 750 feet of sea, with five lines stretching from shore to shore, the number required would be 21,000 of these costly and difficult weapons. The number required annually to maintain such a barrage would also be very heavy, and it is safe to assume that *considerably* over 50,000 mines were employed on the northern barrages alone. From this rough estimate some idea of the work of designing, manufacturing, testing, laying, renewing and watching this one field will be obtained.

There were, of course, in the actual barrage several minefields placed strategically, and probably a far greater number of weapons than that given in the above estimate was needed.

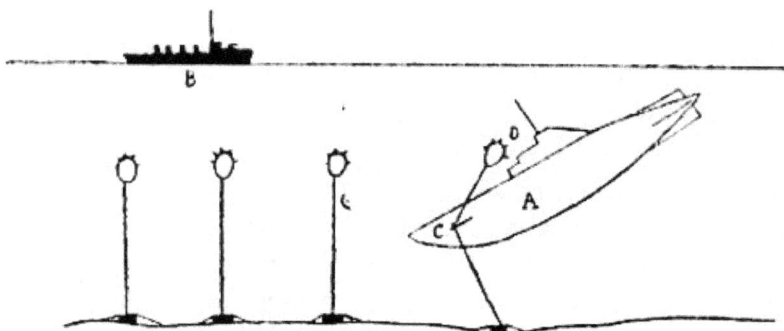

FIG. 31.

Diagram illustrating a mine barrage, or deep-laid mine-field. The submarine A, diving to avoid a surface warship, has become entangled in the mooring of a deep-laid mine which is being dragged down on top of her. These mines are often moored at a depth of 60 feet below the surface, which can then be patrolled by surface warships.

There were also the smaller fields lying between the northern barrage and the one across the Straits of Dover. These were so placed as to catch hostile submarines operating off the east coast of England, or a surface raiding squadron, such as those which in the earlier years of the war bombarded certain British ports.

Finally, when victory had been achieved, there came the cold-blooded task of clearing these immense areas of sea, not only of German mines, laid haphazardly, but also of the thousands of British mines laid methodically and away from neutral traffic.

The English Channel barrage differed from the northern line in several important respects. Being so much shorter (31 miles against 680), it could more easily be made perfect. The swift-running tide, however, greatly increased the difficulty of laying effective mine-fields.

The Lighted Barrage

This southern system consisted, on the surface, of a number of vessels specially built to ride out the heaviest gale at anchor. These were moored at intervals across the Straits of Dover, forming two lines from the English to the French coast. The first line extended from Folkestone to Cape Gris Nez, and the second line about seven miles to the westward of these points (see Fig. 32). Each vessel was fitted with powerful searchlights for use at night, and the dark spaces of sea between were patrolled by large numbers of armed craft.

By this means the only avenues by which hostile submarines could hope to pass on the surface through the barrage at night were the dark lanes of water between the lightships. It was these points which were closely guarded by strong patrol flotillas, whose duty it was to attack submarines attempting to get through and, with the aid of guns and depth charges, to force them to dive below the surface.

Here certain destruction awaited them on the submerged mine-fields. If, however, one line of defence was safely passed by a hostile submarine, there was another to be negotiated seven miles farther on, and once a submarine got between the two

MINESWEEPING GEAR ON A TRAWLER

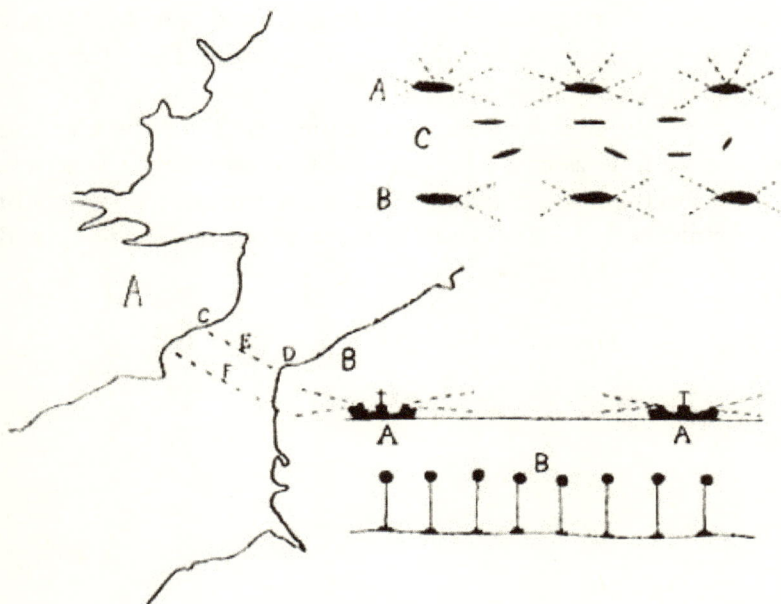

Fig. 32.

Diagram illustrating the Dover lighted barrage. This barrage consisted of two lines of lightships, E and F, from England A to France B. The first line extended from Folkestone C to Cape Gris Nez D. The second line F was situated seven miles westwards of the first line. The small top diagram shows how the two pathways of light, with a numerous patrol between, compelled the U-boats to dive in order to avoid observation and destruction by gun-fire. The lower diagram shows the deep-laid mines arranged to receive the U-boats when they attempted to run the blockade in a submerged condition.

lines her chances of escape were indeed small, for whichever way she turned the surface would be covered with fast patrol craft and at night lighted by the rays of many searchlights, while the under-seas were almost impassable with mines.

If, however, notwithstanding these defensive systems, a submarine succeeded in passing through and getting to work on the lines of communication with the armies in France, there were hydrophone organisations and patrols all down the Channel from the lighted barrage to the Scilly Islands. By this means a U-boat would be seldom out of the hearing of these instruments for more than an hour or so at a time.

The success which attended the perfecting of this vast system was such that German submarines based on the Flanders coast gave up attempting to pass down the English Channel. They

145

tried to go to and from their hunting grounds on the Atlantic trade routes round the north coast of Scotland. Here the great northern systems took their toll.

During the first nine months of the year 1918 the German submarine flotillas at Zeebrugge and Ostend lost thirty vessels, and no less than fifteen of these had, at the time of the signing of the Armistice, been discovered lying wrecked under the lighted barrage.

CHAPTER 14

Off to the Zones of War

Hitherto I have dealt with the scientific training of the personnel, the armament and the general organisation of the antisubmarine fleets, leaving it to the imagination of readers to invest the bare recital of facts with the due amount of romance. If, however, a true understanding of this most modern form of naval war is to be obtained, the human aspect must loom large in future pages.

War, whether it be *on* the sea, *under* the sea, on the land or in the air, is a science in which the human element is of at least equal importance with that of the purely mechanical. It is a science of both "blood and iron."

The armed motor launches described in earlier pages, after being built in Canada to the number of over 500, and engined by the United States, were transported across the Atlantic on the decks of big ocean-going steamships—more than one of which was torpedoed on the voyage. On their arrival in Portsmouth dockyard the guns and depth charges were placed aboard and the vessels thoroughly equipped and fitted out for active service.

Officers and men were drafted from the training establishments of the new navy at Southampton, Portsmouth, Chatham, Greenwich and elsewhere. Each little vessel was given a number, and within a few weeks of their arrival from the building yards on the St Lawrence they sailed in flotillas out past the fortifications of Spithead, *en route* for their respective war bases.

Great secrecy had surrounded the construction of these small but powerful craft, and but few naval men, except those directly engaged in the anti-submarine service, had either seen or heard much of them until they commenced arriving at the different rendezvous.

Among the early flotillas to leave Portsmouth dockyard was one of four ships destined for a base on the east coast of Scotland, and as these speedy little craft raced away north the expectations of both officers and men ran high.

It was in the early summer of 1916, and although the air was crisp, the sea sparkled in the bright sunlight and the sky was a cloudless blue. Only a heavy-beam sea off Flamborough Head had marred the maiden voyage, and they were now on the last hundred miles, with the low-lying Farne Islands fading into the mist astern. By nightfall, if the wind remained light, they would make the Scottish port which was to form their base of operations.

Hitherto these four brand-new little warships, all white wood, grey paint and polished metal, had been plodding over the 600 miles of sea from Portsmouth at what was termed "cruising speed"—a mere 10 knots. The engines had not been opened out to "full ahead" because these delicate pieces of mechanism needed time to settle down to their work before it was safe to drive them to the utmost limit of speed and power, but now that pistons and bearings had been given time to "run in" it was considered safe for the flotilla to increase speed in order to make harbour by nightfall.

A hoist of new, bright-coloured flags fluttered from the squat mast of the leading ship. The steady throbbing of the engines grew suddenly to a low staccato roar. The white waves astern rose up almost level with the counters and clouds of fine spray blew across the decks. This rapid movement through the sun-lit water, with the breeze of passage and the tang of the salt sea in every breath, was exhilarating, and the spirits of those aboard rose with the speed.

Running at nearly half-a-mile a minute, the flotilla forged northwards through clouds of fine, stinging spray, until at a late hour, when the sun was dipping below the horizon and the sea was a sheet of golden light, a smoky line appeared far away to the westward. It was that section of the Scottish coast which in future it would be the duty of these boats to patrol, and as the distance lessened those on board gazed in silence at the gigantic cliffs and black rocks, now tinged with the rays of the dying sun and encircled by the endless ripples which alone broke the peaceful surface of the sea, but one and all were picturing this forbidding coast on the stormy winter nights to come.

Slowly the light faded from the western sky. The cliffs rose up black and sombre, and when the little flotilla turned westwards up the broad waterway leading to the base darkness had closed over land and sea. For some time they picked their way up this sheltered loch. No lights were visible, but more than once a destroyer appeared out of the blackness to make sure of their identity, and each time they were inspected very closely before the guard-ships were satisfied. An armed trawler guided them past dangerous obstructions and then faded into the night. Mile after mile of water was then traversed on courses laid down in confidential orders.

Suddenly a searchlight flashed out from close ahead, followed almost instantly by other blinding rays, which swept the sea for a few seconds, and then all the beams concentrated on the little flotilla, showing up with the clearness of daylight the four low-lying submarine-like hulls gliding speedily through the water. There was a moment's silence, during which the Morse signalling lamps of the M.L.'s were being prepared to flash out their message. A searchlight blinked and there followed another brief interval of silence, then, without warning, a tongue of livid flame stabbed the darkness and a shell whistled overhead. It was followed by other flashes and the sharp reports of quick-firing guns. Columns of water spouted into the air close to the M.L.'s, whose engines had, luckily, ceased to throb. The firing stopped as suddenly as it had commenced. Signals began flashing angrily in many directions. Destroyers tore out of the darkness around into the broad circle of light. Armed trawlers nosed their way in and wicked grey tubes were trained on the now stationary flotilla. Presently the angry flashing of mast head-lights subsided into the regular dot and dash of respectable communication. Several destroyers seemed to be having nasty things said to them, which they answered with a feeble wink, and an armed trawler made futile flashes of explanation.

A little twinkling star, more lofty and dignified than the rest, called up the leading M.L. and was answered with an alacrity that evidently unnerved it, for it flickered and died out. Suddenly it came to life again and winked away at an alarming rate, but all to no purpose, for, true to the old axiom that more haste means less speed, it had to stop and go over the message again, this time sufficiently slow for novices to understand. What it said is a State secret. It is rumoured, however, that several officers were "mentioned in dispatches" for the part they played in this local action, caused by mistaken identity, but alas! their skill and bravery remained unrewarded by an unsympathetic Government.

CHAPTER 15
A Memorable Christmas

No calling tempers the human steel in so short a period as that of the sea. At all times and in every part of the world the sailor-man wages a never-ending fight with Nature in her wildest and most dreaded modes. When to this is added a conflict of nations and their ships, with all the ingenious death-traps of modern naval science, it merely increases the odds against him and serves to steady his hand and brain in order to overcome them.

In a few short weeks the sea had set its stamp on the men of the new navy. Faces became bronzed by the sun, wind and spindrift. Muscles grew hard and eyes and nerves more steady. Each time a vessel went forth on patrol or other duty new difficulties or dangers were met and overcome without advice or assistance, and the confidence of men in themselves and in the ships they worked grew apace.

In many of the principal zones of war, such as the North Sea and the Atlantic, the wind grew colder and the seas more fierce as the short summer passed. Duffel or Arctic clothing was served out to both officers and men. Sea-boots and oilskins became necessary. Balaclava helmets, mufflers and other woollen gear appeared, and men became almost unrecognisable bundles of clothing. The ascent at 4 a.m. from the cabin to the cold, wet deck can be likened only to the first plunge of a cold bathing season. Casualties became more frequent as the enemy intensified his submarine and mining campaign. The news and sight of sudden death no longer blanched the faces of men who knew that it might be their turn at any moment of every day and night. The stir of suppressed excitement when danger threatened no longer manifested itself in every movement, but rather in the cool, deliberate action of self-confidence. In a word, the raw material was being tempered in the furnace of war.

To those in northern seas came the blinding sleet, the slate-grey combers and the innumerable hardships and dangers of winter patrol. A better idea of what these really were will be obtained from the following account of a Christmas spent on a German mine-field.

****

A bitter wind swept the grey wastes of the North Sea and a fine haze of snow, driven by stinging gusts, obscured all except the long hillocks of water which rose and fell around the tiny M.L.—a lonely thirty tons of nautical humanity in as many square leagues of sub-Arctic sea.

Nineteen degrees of frost during the long winter night had flattened the boisterous, foam-capped waves, and now, in the early December dawn, all within vision was of that colourless grey so familiar to those who kept the North Sea on the winter patrol.

It was one bell in the first watch and three shapeless figures clad in duffel coats with big hoods and wearing heavy sea-boots stood silent in the draughty, canvas-screened wheel-house as M.L.822 wallowed northwards through the seas which came in endless succession out of the snowy mist. It was just the ordinary everyday patrol duty, when nothing was expected but anything might happen, so eyes were strained seawards in a vain endeavour to penetrate the icy curtain blowing down from the Pole. Twelve hours more of half-frozen existence stretched in front of these silent watchers, as they clung with stiffened limbs to ropes stretched purposely handy to keep them upright when the little ship lurched more fiercely in a steeper sea.

Of the three figures in the meagre shelter of the wheel-house there was little to distinguish who or what they were, except, perhaps, a cleaner and more yellowish duffel coat and a big white muffler in which the lieutenant-in-command tried, without success, to keep his teeth from chattering and the icy draught from finding its way into the seemingly endless openings of his woollen clothing. What he had been before the Great War and the North Sea claimed him was a mystery to those on board, but the people of more than one capital knew his name. Near by stood a younger man—a boy before the war—who, although pale and dark-eyed, did not appear to feel the intense cold so much, although the dampness of the long-past summer fogs had chilled him to the bone. He was the sub-lieutenant, and hailed from the Great North-West, where Canadian winters had hardened his skin to the stinging dry cold.

The immense bundle of nondescript clothing at the wheel was "Mac," the coxswain, whose voyages in Arctic seas with the Iceland fishing fleet numbered more than his years of life, and whose deep-voiced Gaelic roar could bring the "watch below" on to the cold, wet deck to their action stations in less time than it would take a new recruit to tumble out of his hammock.

Although the silence of the sea seems to settle on its watchers in those northern marches, there was an unduly long absence of comment on the nature of the weather and the prospects of "something exciting" turning up out of the icy mist. The reason lay in the subconscious mind of all on deck, for it was Christmas morning, 1916, and the thoughts of all were dwelling on past years in the cheery surroundings of English and Colonial homes—in vivid contrast to the dismal grey of the North Sea. To break the spell of memory both officers felt would be blasphemy, and yet a feeble attempt at conversation was made every now and then for the sake of appearances.

FIG. 33.
Duffel or Arctic clothing

153

To Mac, from the Orkneys, no such sentiment held sway, for Christmas to him meant little compared with New Year's Day; but this was a special Christmas, for a big plum pudding was being boiled on the petrol stove below, and each roll of the little vessel threatened its useful existence. Eventually he could keep silent no longer and tentatively suggested a change of course to ease the violent lurching. The wheel was spun round with alacrity as the telegraph rang out below and the engines slowed down to a slow pulsating throb. The sharp bows of the patrol boat rose dripping from each green-grey mass of sea as it rolled up out of the white haze ahead and then fell gently back into the trough. The violent pitching gave place to a more easy see-saw movement, and in spite of the cold, which seemed to grow keener every minute to the half-numbed figures on deck, a grunt of satisfaction escaped the helmsman, and visions of steaming plum duff—a present from the Admiral's wife—supplanted the more anxious thoughts of war and the dangers of mine and submarine which lay hidden in the white snow-mists and grey seas around.

The four hands in the forecastle, who formed the watch below, were lying on their bunks, for sitting meant holding on, and were discussing orgies on past Christmas days and planning future ones with a nonchalance bred of daily rubbing shoulders with danger and death. Snatches of popular music hall songs penetrated the closed hatchways, but were drowned by the splash of the sea against the ship's side.

This silent battle with monotony, bitter cold and drenching showers of spray, with several numbing hours on deck, followed by an equal time lying on the bunks below—still cold and wet, for fires and dry clothes were almost unknown in the patrol boats during the long winter months in the cruel northern seas—might have lasted all day, until darkness and increasing cold added their quota to the sum of misery, and the day patrol crept silently into harbour, to be relieved by their brethren of the night guard.

But such was not to be, for it was a Christmas Day that will live for ever in the memory of the men on Patrol Launch No. 822, to be recalled in the peaceful years ahead to eager listeners at many a fireside.

Two bells in the afternoon watch had barely struck when from out of the haze ahead came a low reverberating boom! The three figures on the bridge stiffened to alertness and the chilled blood went coursing more warmly through their veins. A few seconds of strained listening, rewarded only by the noise of the sea, then the telegraph was moved forward, a sharp jangle of bells came from the engine-room and forecastle and the slow pulsating of the motors grew to a loud roar. The watch below came tumbling on to the wet deck, to be lashed with clouds of blinding, stinging spray, which now flew high over the little ship as the 400-horse-power engines drove her at 18 knots through the grey, misty seas.

Experience had made that dull roar familiar to all on board, and it needed no order from the now hard-faced C.O. to cause every man to don his *"capuc"* life-belt in readiness for the hidden dangers which they knew to be strewn in the pathways of the sea ahead.

Mines are moored at a given depth below the surface, usually from six to ten feet. The rise and fall of the tide, therefore, either increases or decreases the stratum of free water above them. This causes these invisible submarine weapons to be more dangerous to shallow-draught vessels, such as motor patrol launches, at low tide, when there is little water between the tops of their horns and the surface, than at high tide. More will, however, be said in a later chapter about mines and the difficulties of laying them.

It so happened that on this occasion the tide was low and the mines consequently extremely dangerous to even the shallowest draught type of warship. The speed of the M.L. was increased until the twin engines were revolving at the rate of 490 a minute.

The snow haze seemed suddenly to grow thicker and all around the flurries of white blotted out the distant view. The minutes of pounding through the slate-grey seas seemed interminably long, and the flying clouds of icy spray stung every exposed part of the human frame.

When about three sea miles had been traversed the engines were stopped and all on board listened for a cry from the sea ahead. The C.O. pulled the peak of his drenched cap farther over his eyes and gazed out into the opaque greyness ahead.

Minutes passed; but little ships cannot rest quietly on the open sea. The lash of the water and the slapping of the meagre rigging drowned any faint sound there might have been, and once more the engines throbbed to the order "Slow ahead!"

Barely had the ship gathered way before a dark object appeared momentarily in the trough of the sea about two degrees on the starboard bow and the next instant seemed swallowed up.

A warning cry from the look-out on the tiny sea-washed fo'c'sle head, a sharp order from the bridge, and, within its own length, the patrol boat swung rapidly to port. At the same moment a dan-buoy splashed overboard to mark the position of the floating mine. A few yards more to the eastward and No. 822 would have appeared in the list of the missing.

Minutes of tense nerve strain followed, for all knew that the ship was in the midst of a mine-field, and the deadly horns which had been momentarily visible on the surface were but a single example of the many which lurked around. Eyes were strained into the grey-green depths, and yet all knew the impossibility of seeing. Again the look-out's warning cry and the engines were reversed, but this time it was not a mine, but the victim of one, holding on to a piece of wreckage.

Willing hands hauled the half-frozen form on board and stanched the blood that still oozed from cuts on the face and neck. Blankets and hot-water bottles were soon forthcoming, and the battered remnant—for both a leg and thigh bone were broken—was placed as carefully as the lurching of the ship would allow in the aft-cabin bunk. Before this could be accomplished, however, a cry again rang out from the watch on the fo'c'sle head and yet another body was hauled aboard, but the shock or the cold had here taken its toll.

The sea around was searched in vain for further survivors. A few planks, a signal locker, a broken life-raft and a meat-safe were all that was left of the trawler *Mayflower*, homeward bound from Iceland to Grimsby.

A silence seemed to brood over the patrol boat as she slowly picked her way out of the mine-field. The crew went about their tasks without the usual jests and snatches of song, and the pudding, which but a few short hours before had seemed the most important event of the day, lay unheeded on the floor of the galley, where it had been thrown by the cook in the haste for hot water.

In the failing light of the December afternoon the bow of the patrol boat was turned shorewards, and, with a rising sea curling up astern, she raced through the slate-grey water with her burden of living and dead. It was one of those moments which call for a rapid decision on a difficult point, when the order had to be given for the course to be laid for harbour, and the C.O., cold and miserably wet after seven hours on the bridge, wore an anxious look. He knew not which had the greater claim, the desperately wounded man in the cabin or other ships which might bear down on the mine-field during the long bitter night. It was a point on which the rules of war and the dictates of humanity clashed.

Again the ship was turned into the rapidly darkening east, and all through that bitter night the field of death was guarded. Stiffened fingers flashed out the warning signal when black hulls loomed out of the darkness. Numbed limbs clung for dear life when green seas washed the tiny decks, and when dawn broke over the waste of tumbling sea the men on M.L.822 knew that Christmas Day, 1916, would live for ever in their memory.

CHAPTER 16

The Derelict

There are few things more heart-breaking than sea patrol, which forms the principal duty of anti-submarine fleets. Hours, days and even months may pass with nothing to relieve the monotony of grey sea and sky, with occasional glimpses of wave-tossed ships.

There are, of course, intervening periods in harbour, when fierce gales howl overhead, and guard duty on rain-swept quaysides, or sentry-go in blinding snowstorms, comes almost as a relief from the sameness of winter days on northern seas.

It is, however, the unexpected which generally occurs in war, and during those terrible winters from 1914-1918 it was the ever-present hope of action that kept the spirits of many a sailorman from sinking below the Plimsoll line of health.

Sometimes the happenings were grave and at other times gay, but always they were welcomed eagerly, as providing excitement or change, with something to talk about in the unknown number of dreary weeks ahead.

An episode of this kind occurred one snowy January night in 1917 on the quayside of a northern seaport. The commanding officer of one of the patrol boats in the harbour was going ashore to stay for the night with some friends. Knowing that his ship was due to proceed to sea early the following morning, he took the precaution to place a small alarm clock in the big pocket of his bridge-coat. Groping his way in the darkness and blinding snow across the gangway leading from the ship to the quay, he succeeded in reaching the dock wall. Almost instantly he was challenged by a military sentry on duty and was about to reply when a loud buzzing noise came from his pocket. He had not thought of ascertaining at what time the alarm clock had been set for and the consequences were distinctly unpleasant.

The sentry, hearing the curious buzzing sound coming from the darkness directly he had given the challenge, and thinking it came from some form of bomb, lunged smartly with his bayonet at the spot from which the sound emanated.

Fortunately the officer was near the edge of the dock wall and did not receive the full effect of the thrust. The bayonet tore his coat and pushed him violently over the edge into the icy water of the harbour. His lusty shouts caused searchlights to be turned on and he was rescued promptly, but the episode, small and unimportant as it was, caused considerable merriment—except to the principal actor—for many days afterwards.

All this may sound much like heresy to those who think that naval war means constant fighting, with all the pomp and circumstance of old-time battles. There are, it is true, never-to-be-forgotten moments when the blood surges and pulses beat rapidly, when the months of weary waiting are atoned for in as many minutes of swift action. Such were Jutland, Zeebrugge, Heligoland, the Falklands and many an unrecorded fight on England's sea frontier in the years just past. Such times pass rapidly, however; they are the milestones of war, leaving the weary leagues between, in which there is so much that is sordid and even ghastly, as will be seen from the following.

****

The sea offers but few sights more melancholy than the wave-washed derelict—the now desolate, helpless and forlorn thing that was once a *ship*, the home of men—seen in the half-light of a winter dawn, rising and falling sluggishly on the dirty grey swell—the aftermath of storm—with white water washing through its broken bulwarks, yards and sails adrift, a thing without life on the sad sea waves.

A wireless message from a ship passing the derelict on the previous day had brought an M.L. from the nearest naval base to search the area, and after a night of wandering over shadowy grey slopes of water the dawn had revealed it less than two miles distant.

There could be no doubt as to its nationality, for the white cross of Denmark, on the red ground, was painted on the weather-beaten sides, now showing just above the sea. Deserted and half-waterlogged, it was being kept afloat by a cargo of timber, some of which could be seen in chaos on the deck.

The M.L. approached cautiously, with thick rope fenders over her rubbing-streak to prevent the frail hull from being damaged. This coming alongside other ships in the open sea, except in the very calmest of weather, is a ticklish manoeuvre, and requires considerable skill in the handling of these small and very fragile craft. What would be considered quite a light blow on the stout hull of any ordinary ship would crush in the thin timbers of a patrol launch, for in the construction of these boats speed and shallow draught were the predominant factors considered.

When the M.L. had been made fast on the lee-side of the derelict a boarding party scrambled over the damaged bulwarks on to the sea-washed deck. Here was a scene of chaos—rigging tangled and swinging loosely from masts and yards; sails torn and shreds still clinging to ropes and spars; loose planks of her deck cargo lying all over the place, and a general air of abandon and desolation difficult to describe.

A mass of broken woodwork in the well of the ship was soon discovered to be the remains of a deck-house, and this gave the first clue to the reason for her sorry plight. Pieces of shrapnel were found sticking in the timbers, and further search revealed shell-holes through the hull and cut rigging. A signal was flying from the mizzen halyards, and the name on the counter, although spattered with shot, was still, in part, decipherable— *Rickivik, Copenhafen.*

So the officer in charge of the boarding party commenced his report with the name of the ship and the port from which she hailed, adding thereto the evident fact that she had been heavily shelled—just a brief statement which left to the imagination all the incidents and, alas! tragedies of an unequal fight.

A high-explosive shell had struck the little raised poop, demolishing the hatchway leading to the cabins beneath, and some heavy work with axe and saw would have been necessary to obtain an entry had an easier way not been available through the shattered skylight. In the low-roofed cabin all was disorder. Tables and lockers were smashed, and the shell which had burst overhead had filled the place with heavy broken timbers from the deck above.

So low was the cabin roof of this small three-masted barque, and so dark the interior, that it was difficult to see about. A lantern was procured and a careful search commenced. The yellow light fell on drawers pulled out and their contents—when worthless—flung on the floor; glasses and bottles smashed and a quaint old China figure lying intact on the broken timbers. But of the ship's papers there was no trace, with the single exception of an old Bill of Health, issued six years previously in Baltimore. Then the area of search moved from the cupboards and drawers to the floor—broken by a shell which had evidently penetrated the ship's stern and passed longitudinally through the cabin, exploding near the base of the companion-hatch.

Presently a startled exclamation, followed by a call for the light, came from the gloom around the stairway. Two of the boarding party searching among the debris had stumbled across something which, instinctively, sent a cold shiver through them. The light, when moved in that direction, dimly revealed the body of a man lying face downwards on the floor. Only the lower half of the figure was, however, visible, a mass of shattered timbers having collapsed on the head and shoulders. That life had been extinct for some considerable time was evidenced by the sickly odour which hung heavily in the less ventilated parts of the cabin, and the work of extricating the body was not commenced before the whole ship had been searched for possible survivors.

This work occupied a considerable time, but nothing of importance was discovered until a slight noise, not unlike the feeble, inarticulate cry of a child in pain, came through the timbers from some distant part of the hold. It was repeated several times, and the sailors, without waiting for orders, set hastily to work to find out the cause.

The hatches were carefully removed, but only floating timber could be seen. Then the sound came again. This time it was unmistakable and relieved the tension. A little grim laugh from the searchers was followed by much poking about with a long piece of wood on the surface of the flooded hold under the decking, and some minutes later a large pile of timber floated into the light from the open hatchway, supporting a big tortoiseshell cat, looking very wet and emaciated. "Ricky"—for such is her name now—proved to be the only living thing on that ill-fated ship.

The boarding party returned to the cabin and commenced the objectionable task of extricating the dead body from the mass of wreckage. The work proceeded slowly, for the heavy broken timbers pressed mercilessly on the object beneath, and when at last it lay revealed in the dim lantern light its ghastly appearance caused all to step back in horror. It was a headless corpse!

CHAPTER 17

Mined-In

How many people realise that, with a single unimportant exception, there was no part of the English or Scottish coast which was not mined-in at least once by German submarines during 1914-1918? Harbour entrances, often less than two miles from the shore, were repeatedly blocked by lines of hostile mines, laid by U-C boats through their stern tubes, in which they seldom carried less than fifteen to twenty of these deadly weapons, without the vessels rising to the surface either when approaching the coast, laying the mines or effecting their escape.

Many important waterways, such as the Straits of Dover, the mouth of the Thames, the approaches to Liverpool, the Firth of Forth, Aberdeen, Lowestoft and Portsmouth, were repeatedly chosen for this form of submarine attack. At one base alone no less than 400 mines were destroyed by the attached anti-submarine flotillas in one year, and round the coasts of the United Kingdom an average of about 3000 of these invisible weapons were located and destroyed annually.

What this meant to the 24,000,000 tons of mercantile shipping passing to and fro through the danger zone *every month* will be better realised when it is stated that less than 400 merchant ships were blown up by mines during the three years of intensive submarine warfare.

The losses among the minesweeping and patrol flotillas, which were mainly responsible for the crushing defeat of this piratical campaign, were, however, very heavy. They amounted to over 200 ships and several thousand men. Few will therefore deny to those who lived and to those who died a share in the glory of the great victory.

Statistics make but uninteresting reading, and from the following account of what happened off a big Scottish seaport while the inhabitants ashore slept in peace and safety a better idea will be obtained of the arduous nature of the work of minesweeping and patrol in time of war than could possibly be imparted by pages of figures.

****

The early dusk of a winter evening was settling over a white land and a leaden sea. A mist of sliding snow increased the gloom and blotted out the vessels ahead and astern as the line of patrol boats left the comparative warmth and security of one of the largest northern harbours for twelve hours in the bitter frost on night patrol.

The cold was intense and of that penetrating nature which causes men to shiver even in the thickest of clothing. Although some eighteen degrees of frost had flattened the sea, a freezing spray still blew in showers over the narrow deck and, for just a few minutes, the lead-grey sky gleamed dully red as the sun dipped below the snow-covered land.

The crew of the M.L. moved about the cramped deck stiffly, for they were clad in duffel suits, oilskins and sea-boots, and little but their eyes and hands were visible. The officer on the small canvas-screened bridge was likewise an almost unrecognisable bundle of yellow and white wool and black leather. As a contrast, however, to the whitening deck and snow-clad men, the reflection of a warm yellow light came up through the wardroom hatchway, and more than one longing glance was cast down into the snug interior.

These men were not all hardened by long and severe sea training; many of them formed part of the new navy, gaining experience amid the bitter cold and dangers of the grey North Sea. A call for the signalman came from the bridge, and a boy, who had been swinging his arms to warm his numbed fingers, responded smartly. The lieutenant-in-command wiped the snow from his eyes as he peered round the canvas side-screen and asked tersely what the next ship ahead was trying to signal.

The boy seized his semaphore flags and went out on to the spray-swept fore-deck, steadying himself against the fo'c'sle hatch cover. He flinched at first when the spray stung the exposed parts of his body, and then, with straining eyes and dripping oilskins, he managed, after the words had been repeated several times, to read the signal which was being sent down the line from the leading ship somewhere in the white haze ahead.

"Proceed independently to allotted stations for night patrol" was the order then conveyed to the bridge and afterwards passed on by flag to the next astern. When the last ship had received the signal each unit of the flotilla swung out of line and disappeared in the sliding snow.

As the darkness increased the cold strengthened and a little bitter wind began to moan through the scanty rigging. Men stamped their feet and swung their arms to increase the circulation in numbed limbs, and every now and then during the next three hours one member of the watch on deck would disappear for a few minutes down the galley hatchway to drink a cup of hot cocoa, which, so far, the cook had succeeded in keeping warm on the ill-natured petrol stove.

At 9 p.m. the first watch was over and half-frozen men climbed stiffly down the iron ladder into the tiny fo'c'sle, where the heat and fug of oil stoves caused their thawing limbs to throb painfully. The starboard watch, fresh from the heat of the tiny cabin, whose four hours on deck now commenced, were shivering in the icy wind and showers of spray.

Glancing at the dimly lit chart on the small table cunningly fitted into the front of the wheel-house, the commander noted the approximate position of the ship in the 140,000 square miles of sea and snow around, and then turning to the coxswain, whose "trick" it was at the wheel, he gave the necessary orders for the course and speed. The duty of this vessel was to patrol certain approaches to the great harbour on which the flotilla was based until relieved at daybreak by another unit, and, as merchant ships had many times been attacked in these waters, a sharp look-out was necessary. To carry this out effectively in the darkness and driving snow was a task calling for all the qualities of dogged endurance inherent in the British sailor.

For over two hours nothing was seen or heard except the moaning of the wind and the lash of the sea, but shortly after midnight one of the look-outs reported the sound of engines away to the starboard.

The M.L.'s propellers were stopped and the watch on deck listened intently. The splash of the sea and the many noises of a rolling ship drowned any other sound there might have been, and the patrol was then continued. Less than half-an-hour later, however, the clank! clank! clank! of engines again became suddenly audible, and the vessel was turned in the direction of the sound.

The engines were put to full speed ahead, and as each comber struck the bows the little ship trembled from stem to stern, and clouds of icy spray swept high over the mast. The big steel hull of some man-o'-war or merchantman might suddenly loom up out of the darkness so close ahead that no skill could avoid a collision, and the eyes of all aboard were gazing alertly into the blackness of the night.

Five minutes' dash through the blinding, stinging spray and the engines were once more shut off to listen. The curious clanking noise had, however, ceased, and although hydrophones were used to again locate the sound, there was no result, only the ceaseless wash of the sea and the low moaning of the wind. Another mile or so of pounding through the waves, followed by an interval of listening, brought the same discouraging result, and the slow, monotonous routine of patrol was continued.

The stinging frost of the night became the numbing cold of early morning, and the long hours in the snow and icy spray had left their mark on all. Limbs were stiff and sore. The edges of wet and half-frozen sleeves rasped swollen wrists. Faces smarted and eyes ached, but little was said in the way of complaint, for men grow hard on northern seas or else succumb to the hardships.

When the first dim light of a winter dawn broke reluctantly over the grey tumbling sea and whirling snow another night patrol was over, and the cheering thought came to all that soon the welcome warmth and shelter of club and recreation room would embrace them for the brief hours of daylight, while others kept watch upon the seas.

It had been snowing hard for the past twenty-four hours, but as the light of a new day strengthened it eased somewhat, and away to the westward the blue outline of the land became visible. The fitful wind of the night rose to a stiff breeze, but no one paid much attention to the increasing volume of bitter spray which swept the deck as the grey-green rollers put on their white caps of foam, for the ship was heading towards the harbour and their vigil was over until darkness again closed down.

Few things are more trying to the temper than to be kept waiting for relief after a bad spell at sea, and but few crimes are more heinous than to leave the watched area before another patrol takes up the never-ceasing duties. Therefore, if peace and quietness and an absence of insulting signals counted for anything, it ill behove any ship in the day patrol to keep her opposite member of the night guard waiting.

This time the relief was late and the M.L. steamed angrily up and down, with all eyes strained shorewards. Then the first of the line of armed trawlers and motor launches crawled out of the harbour in a smother of black smoke. When barely half-a-mile of sea separated the incoming and outgoing ships a loud reverberating boom rolled over the sea. So great was the explosion that the shock of it was felt rather than heard, and a gigantic column of black smoke, rising over 100 feet into the air, appeared to engulf the leading unit of the trawler patrol.

Regardless of the danger, the C.O. of the motor launch sent his swift shallow-draught boat flying over the mine-field into the floating debris. The only two mangled survivors had, however, been picked up by the trawler astern of the ill-fated vessel, which had been literally blown to pieces, nothing remaining afloat when the smoke cleared away except a signal locker and a few timbers.

More than one of the other vessels, whose engines had been stopped immediately the explosion occurred, narrowly escaped drifting down with the tide on to the field of hidden mines, but

A MOTOR LAUNCH OF THE NAVAL PATROL

with the skill and presence of mind gained by similar experiences in the past both the trawler unit and the M.L. flotilla were extricated without further loss.

It was evident from the fact that several of the mines were barely submerged and could be dimly seen from the decks that the work of laying them had been done hastily under the cover of night, and a sense of keen sorrow and disappointment pervaded the vessels of the night guard. Once again climatic conditions had favoured the enemy. In those long winter hours of impenetrable blackness and driving snow no watch, however efficient, could be relied upon to prevent such operations from being occasionally carried out. It was merely the chance of war, but nevertheless it was felt keenly, and the sense of responsibility was not dispelled until some weeks later.

When the *sweepers* arrived it was soon discovered that the harbour was temporarily mined-in. Signals were exchanged with the "Senior Naval Officer" of the base, and the night guard was ordered to assist in preventing shipping from attempting to enter the harbour before the approaches had been swept clear and the mines destroyed. Weary ships with disappointed crews once more turned seawards, but the physical discomforts of stinging spray and frequent snowstorms passed almost unnoticed in the efforts of the flotilla to prevent the ceaseless stream of ocean traffic from approaching the danger zone unnoticed in the blinding white haze.

Tired limbs were forced to continued efforts and numbed faculties were goaded afresh. Big ships loomed out of the mists around and were informed of the dangers and directed into the pathways of safety. Trawlers returning from the fishing-grounds of the far north had to be intercepted, local craft piloted round the mine-field in the shallow water close inshore, signals flashed to the outer patrols, and the hours of daylight and activity passed quickly by.

By seven bells in the afternoon watch the dusk of the long winter night began again to settle over the sea, blotting out one patrol from another. On this as on many other similar nights spent in the bitter frost, thick sea fog or flying spume, in waters infested with mines and hostile submarines, certain senses became dulled, though the brain remained alert and the limbs as active as cramp and cold would allow. But the little incidents of those long hours are lost in blurred memories of cries from the look-out, hulls towering out of the blackness, the flashing of Morse lamps, the ceaseless and violent pitching and rolling of a small ship, moments of tense excitement, followed by hours of cold and an utter weariness of the soul.

When the first pale streaks of returning daylight had turned to the fiery red of a frosty sunrise, dirty and unshaven men moved painfully about the slippery decks. The sea had flattened in the night and the snowing had ceased, but twenty degrees of frost had gripped the wet decks and the soaked clothing. As the vessels stood towards the shore weary eyes were turned anxiously on the signal station, but not yet was the recall to be hoisted, for although the seas around had been swept clear of mines, there was still a careful inspection to be made before the area could be reported clear, so that ships might come and go.

When at last a line of flags fluttered to the distant mast-head away on the hill ashore, and the signal-boy read out, "M.L.'s to return to harbour," there was a feeble cheer.

****

On a calm, frosty morning some three weeks later the boats of the old night guard, now doing their spell of day duty, discovered a long trail of thick greenish-black oil on the surface leading seawards. It was evident that a hostile submarine had rested during the previous night on the sandy bottom in the shallow water close inshore and, rising to the surface, had made off at daybreak. The trail was followed and information was quickly received from an Iceland trawler, which had passed the submarine on the surface some two hours previous. Ships were concentrated by wireless, and although it did not fall to the lot of the M.L.'s to give the *coup de grâce*, they had the satisfaction of returning to harbour with the knowledge that their honour had been retrieved, and yet another German submarine would never again commit outrage on the high seas.

CHAPTER 18

The Casualty

There were duties performed by the new navy which bore no relationship to anti-submarine fighting, or, in fact, to warfare at all, unless it was to the ceaseless battle waged between all who go down to the sea in ships and the elements they seek to master.

One such as this occurred at a little northern seaport in the late winter of 1917, unimportant and scarcely worth relating except as an illustration of the diverse services rendered by men of this great force during the years of national peril.

****

The gale was at the height of its fury when the March day drew to a close. The whole east coast of Scotland, from John O'Groats to the mouth of the Tweed, was a study in black and white—the white of foam and the black of rocks. All the mine-sweepers and smaller patrol ships had been confined to their respective bases for several days, and in a certain small harbour many of the officers and crews of the imprisoned ships were spending their time ashore, in the warmth and cheery comfort of hospitable firesides. The boisterous day became a wild night. The wind howled and whistled over the barren moors and through the streets of the small fishing town. Houses trembled and chimneys rocked under the blasts. Although a watch on the signal tower and elsewhere was religiously maintained, it was of little value, as all that could be seen in the darkness to seawards was a hazy mist of flying spray which the wind whisked from the surface and carried several miles inland.

Standing back from the sea, and some half-mile from the centre of the little fishing town, stood a substantially built house, more commodious and better furnished than many of its neighbours, which had providentially fallen into the temporary grasp of one of the married officers of the patrol flotilla, who generously kept open house for his less fortunate brothers-in-arms.

On this wild winter night the interior looked excessively cosy and inviting. Before a big blazing fire of logs sat three officers, talking between copious sips of whisky and soda. Their conversation was subdued and their inhalations of cigar smoke long. By their side were the faithful women who had followed them from the comforts of home and the gaieties of the great southern cities to this remote corner of northern Scotland. They too were talking among themselves and knitting for the crews of their husbands' ships.

This quiet domestic scene would have gone on uninterruptedly until a late hour, for it was seldom that such precious moments of rest and contentment could be snatched amid the ever-recurring duties and the turmoil of war, had it not been for one of the officers who glanced ruefully at his wrist watch and then apologetically informed his host that it was his turn for night duty on the signal tower.

Scarcely had he risen from the fire and moved towards the door of the room, however, before the dull boom of a gun was borne on the howling wind. All stood still and listened. The women ceased their knitting and looked up apprehensively. Then a minute or so later the boom came again, this time in a lull of the storm, and it sounded nearer.

The three officers hurried into the hall to get on oilskins and sea-boots, but almost before this could be done there came a report which echoed sharply through the little town. They knew the sound only too well, for the coast was a dangerous one. It was the reply of the life-boat crew to the call of distress, and with one accord they moved towards the door. Almost instantly it was thrown violently open and the rush of wind and rain extinguished the hall light. For the next few minutes they were struggling against the gale, battling their way to the lofty little signal station, impeded in every movement by driving rain, flying scud, intense blackness and flapping oilskins.

When they had reached the coast and mounted the rough stone steps leading to the elevated look-out tower, a clear sweep of the dark, foam-crested surface was obtained, and the news was shouted above the roar of the gale that somewhere out in the night, amid the tormented waters, a ship was in distress, though the flying spray made it impossible to locate the exact direction.

Below the signal tower, and built on a mass of rock projecting into the half-sheltered water inside the concrete pier, was the life-boat house. From this point the white rays of a chemical flare lighted up the surface of the sea as far as the harbour bar, which, with its flanking rocks, resembled a seething cauldron. Into this the life-boat plunged from its inclined slipway, and was almost instantly swallowed up in the outer ring of darkness and spray. The flare died out suddenly and the night seemed even blacker than before.

After a brief struggle with the wind, now blowing at a speed of over seventy miles an hour, the men who had assembled around the signal station made their way out on to the spray-swept breakwater, and there waited for the coloured rocket from the life-boat which would signify that she had found the wreck.

Nearly an hour passed but no sign came from the darkness and boiling sea. Then a light appeared momentarily on the harbour bar and was lost in the smother of white. A few minutes later a grinding crash came from the rocks less than a hundred yards distant from the end of the breakwater.

The groups of sailors standing under the lee of the wall, chafing at their apparent helplessness and gazing anxiously out to sea, were suddenly electrified into action by a few sharp orders from the oilskinned commander. A minute or two of seemingly inextricable confusion resulted in the beams of a portable searchlight flashing out from the spray-swept breakwater and lighting up rocks, foam, and a big three-masted Norwegian sailing ship, with sails torn, her fore-mast broken off short and every sea lifting high her stern and driving her farther on to the half-hidden tongues of stone. Even as the light played on her she heeled over to starboard at an angle of about forty-five degrees with an ominous rending of timbers which sounded above the roar of wind and surf.

Orders were bellowed through a megaphone, and again men moved quickly in all directions. This time a fiery rocket, bearing a life-line, soared from its tube with a loud hiss and sped across the hundred yards of boiling sea. It straddled the wreck. The thin line it carried was soon exchanged for a stout hawser—hauled from the breakwater—and this was made fast to the stump of the mainmast, which had followed the other "sticks" overboard when the vessel heeled over on the rocks. It was now floating, wrestling and tugging at the mass of confused rigging, and pounding dangerously at the ship's side.

One by one the unfortunate Norse crew were hauled over the harbour bar in the breeches-buoy by fifty willing British sailors, and the first to come was the captain's wife and little daughter.

There was but one casualty, and that among the rescuers. The stretcher was lifted from the ambulance at the door of the substantially built house standing back from the little town. A white-faced woman ran out into the storm. She had spent a year of nights and days half expecting such as this, and now that it had come the blood seemed to ebb from her body, and at first she scarcely heard a familiar voice assuring her that it was only a cut on the head from a broken wire rope.

CHAPTER 19

How H.M. Trawler No. 6
Lost Her Refit

An earlier chapter described the periodical overhauls necessary to keep the ships of the hard-worked auxiliary navy in
proper fighting condition. These "refits" were needed not only
by the ships but also by the men who worked them. They came
about once a year and lasted for two or three weeks, during
which time the crews were able to go home for at least a few
days of much-needed rest.

To describe how everyone, from commander to signal-boy,
looked forward to these spells of leave is unnecessary. Let the
reader imagine how he himself would feel after nine or ten
months of the monotony and danger, to say nothing of the
hardships, of life at sea in time of war.

There was, however, another consideration, one seldom referred to but nevertheless unavoidably present in the minds of
all. Each time a refit came round there were ships which would
never be docked again, and comrades who had missed their
leave. Men told themselves that the luck they had enjoyed for so
long could not last, and it is about one of these, in a fight against
overwhelming odds, that the following story deals.

Three of his Majesty's armed trawlers were plunging through
the sea on their lonely beat in the Western Ocean. The Hebrides
lay far to the southward, and less than two days' steam ahead lay
the Arctic Circle. These cheerless surroundings, however, found
no echo in the hearts of the watch below on the leading ship of
the unit, who were lounging on the settees in the oil-smelling

fo'c'sle discussing their prospects of long leave, for their ship was to "blow-down" for a thorough refit when they returned to harbour in less than three weeks' time.

On the deck of the same vessel two officers, standing in the shelter of the wheel-house, were sweating and shivering in patches, but also happy with the thought of the forthcoming reunion with their families and the brief enjoyment of the comforts of home after seven long winter months' wandering, with soul-destroying monotony, over the windswept wastes of England's frontier. The watch on deck, with the exception of the helmsman and look-out, crouched under the lee of the iron superstructure, alternately swinging their arms and stamping their heavily booted feet, but they too were mentally impervious to the dismal surroundings.

Of the second ship in the line the same cheery story cannot be told. She was jealous of the first. It would be another two months at least before she would go in dock for refit; and among the watch below there were three new hands on their first voyage, two of whom would, just then, have preferred the peace and stillness of the sea bottom to the friskiness of the surface.

The third trawler was a happy little ship, for although the junior of the unit she had been very fortunate in securing a "Fritz" all to her own cheek less than three months before.

This, then, was one of the units on the Outer Hebrides and Iceland patrol during the winter of 1915, and they seemed to be the sole occupants of the leagues of water around.

It was barely eleven o'clock, Greenwich time, when they reached the last ten miles of their beat, and speed was reduced so that they would not have to turn about and begin steaming back over the course they had come until the morning watch went below at midday. This was an artful though harmless arrangement to enable those going off duty to have a meal and at least an hour's rest in peace, as on the voyage back both wind and sea would be astern and the vicious lurching of the small ship reduced to a minimum.

The time passed slowly, as it generally did on patrol when nothing exciting was afoot, but a few minutes before the awaited eight bells the officer on duty snatched up the binoculars, and almost simultaneously the look-out gave a warning shout which caused the attention of everyone on deck to suddenly become strained.

Away to port, less than half-a-mile distant, the thin grey tube of a periscope could be seen planing through the waves, with a fringe of white foam blowing from its base. There was a hoarse cry down the fo'c'sle hatch for "All hands on deck!" The telegraph tinkled for "Full ahead!" A signal was made to the ships astern for concerted action. The gun was manned, and the leading trawler, now cleared for action, headed towards her underwater opponent.

The other two vessels of the unit put on speed and spread out until all three were line-abreast and about two cables apart. In this formation the chase was maintained for some twenty minutes, when a second submarine appeared above the surface away to starboard. She appeared to be a large vessel and would probably have turned the scale at 1000 tons.

It was at this early stage in the action that the mistake was made. The leading trawler immediately opened fire, but the range was considerable and the shells fell short. Signalling to the other two trawlers to continue the chase of the first submarine sighted, she headed straight for the largest of the two hostile craft to engage her at close range.

While this was in progress the first submarine came to the surface and proved to be also a larger craft than had been anticipated. The two trawlers chasing her immediately opened fire, but her superior surface speed soon placed her out of range of the comparatively small guns then carried by the trawler patrols.

Now came the surprise. Almost simultaneously the two submarines opened fire from heavy guns. The shells at first fell wide, but in a moment the British officers realised that they were outranged, for whereas their shells were falling short, those from the enemy whistled over their heads and ploughed up columns of white water over a cable's length astern.

To increase speed and so reduce the range became imperative, and the steam-pressure in the trawlers' boilers was raised to bursting point by the simple expedient of screwing down the safety valve. For some minutes it looked as though the effort would be successful, and then the range slowly increased again and "short" after "short" was registered by the gunners.

At this psychological moment a German shell carried away the funnel of the leading trawler and smothered her decks with smoke. When a temporary shield had been rigged it was observed that one of the other patrol ships had been crippled by a direct hit and was in a sinking condition.

It now became evident that the superior speed and gun-power of the submarines enabled them to keep out of range of the trawlers' weapons and to ply their long-range fire with telling effect.

The officer in command of the patrol at once realised the mistake he had made when opening the action, in betraying the power of his own guns before he was sufficiently close to the enemy to ensure hits, and he cursed this want of foresight which looked like costing the life of the flotilla. Given one direct hit on each of his two powerful opponents and they would in all probability have been put out of action, but instead he had only the mortification of seeing every shell fired fall short, while his own vessels were being battered to pieces by the long-range guns of an enemy with whom he could not close.

The withholding of fire while hostile shells are bursting around is one of the many severe strains imposed on the human mind by modern war, and in anti-submarine tactics it often means the difference between victory and defeat, which, followed to its logical conclusion, is generally life or death.

One hope now remained—that by skilful manoeuvring the trawlers could be kept afloat until help arrived; but in those wastes of sea no vessel might pass for many hours, and even then not a warship.

Such is the working of Fate: the leading trawler of the unit was to have been fitted with wireless while under the approaching refit, and with its aid patrol cruisers or fast destroyers could soon have been brought to the scene of operations.

Thirty minutes later the crippled ship, the junior member, gave three defiant shrieks with her siren and slid under the surface with her colours flying. For over two hours the others manoeuvred to get one on each side of the submarines to enable them to get the few shells remaining in their magazines home on the target, but so great was the disparity of both range and speed that at five in the evening nearly half their crews were dead or wounded, and a little while later the ice-cold water closed over the leading ship. Still the other fought on, but as dusk closed over the sea she too went down in this obscure fight.

No search for possible survivors was made by the submarines, which glided westwards into the smoky red afterglow, leaving the bitter cold to finish the work of death.

****

A big armed liner of the Tenth Cruiser Squadron had heard the distant firing and came upon the scene just before darkness finally closed over. Four bodies were still lashed to a raft, but in all except one life was extinct.

When the doctors bent over the half-frozen form in which a flicker still lingered they shook their heads. Death waged a stern battle even for this last relic, but life triumphed, and when the agony of returning animation had ceased the sole survivor told the cruiser's mess how Trawler No. 1 had lost her refit.

CHAPTER 20

The Raider

Everyone familiar with English history knows that it was a severe gale which destroyed the scattered and defeated units of the Spanish Armada in 1588, and that, in more modern times, it was the coming of darkness which prevented the British Grand Fleet from turning the victory of Jutland into a decisive rout. Such historical examples of the effect of the weather, and even ordinary climatic changes, on the course of naval operations could be multiplied almost indefinitely. Not only are the movements of the barometer important factors to be considered in the major operations of naval war but also in minor sea fights.

Comparatively few people are, however, aware that one of the largest and most destructive of German mine-fields was laid off the British coast during the Great War by a surface ship which escaped detection through darkness and storm.

****

The barometer had fallen rapidly, and clouds rolled up from the north-west in ragged grey banks which scudded ominously over a cold steely blue sky. For some days the sea had been moderately calm, but it was mid-winter and quiescence of the elements could not be expected to last. Slowly the face of the Atlantic grew lined with white. It began with a moaning wind which soon developed into a stiff gale, accompanied by heavy storms of sleet and snow.

One of his Majesty's ships coming up the west coast of Ireland found herself heading into the teeth of the gale. As the afternoon wore on the wind increased in violence and the ship rolled and plunged heavily, smothering herself in clouds of flying spume. The driving sleet made it difficult to see more than a cable's length in any direction, and when dusk closed over the storm-swept ocean the ship was headed for a sheltered stretch of water close inshore.

Every stay and shroud whistled its own tune as the gale roared past. Foam-crested waves hurled themselves in a white fury against the plunging, dripping sides, piling up on the port bow and racing aft in cataracts of water which threatened instant death to any luckless sailor caught in their embrace. The lashings on the movable furniture of the decks, although of stout rope, were snapped like spun-yarn, and much-prized, newly painted ventilators, boat-covers, fenders, deck-rails and other necessary adornments were swept overboard by the ugly rushes of green sea. The iron superstructure and bridge-supports resounded to the heavy blows of the water, and the ship trembled as she rose after each ghastly plunge.

The blasts of wind which struck the vessel with increasing violence had swept unimpeded over 5000 miles of ocean and carried in their breath the edge of the Arctic frost. The sleet felt warm compared with it, and the flying spray lost its sting.

The forty-eight sea miles lying between the ship and the sheltered strait seemed endless leagues, for the speed had to be considerably reduced to avoid serious damage from Neptune's guns. The minutes of twilight grew rapidly less, and with the coming of darkness a new danger threatened. The ship was approaching a rock-strewn coast with no friendly lights to guide her, and every now and then lofty masses of black stone rose up, dimly, from their beds of foam. It was an anxious half-hour, and ears were strained for the warning thunder from surf-beaten rocks which sounded at intervals even above the roar of the gale.

Fortunately the entrance to the sheltered waterway was broad, and almost before it could be realised the sea grew calm. Although the wind still shrieked and moaned, the waves rose barely three feet high. Great cliffs, invisible in the darkness and driving sleet, protected the strait, and as the vessel picked her way to a safe anchorage closer under the lee of the land the wind lost its giant strength and the howling receded into the upper air.

Throughout the night the comparatively small warship rode safely at anchor, innocent of what was taking place out in the blackness and the storm. When morning broke the gale had lost some of its force, and streams of pale watery sunlight shone between the low-flying clouds on to a boisterous sea.

****

Running before the wind and sea the German raider *Frederick*, carefully disguised and loaded with several hundred mines, approached the British coast. The gale was increasing in force as darkness closed down, and heavy showers of sleet shielded her from the view of any passing craft. The weather was ideal for her dark purpose, which was to lay a mine-field over a stretch of sea where it was thought the Anglo-American trade routes converged.

For the first few days out from Wilhelmshaven the weather had been misty with heavy snowfalls, conditions enabling the mine-layer (and afterwards raider) to run the blockade and elude the network of patrols, not, however, without some very close shaves. On one occasion a large auxiliary cruiser passed in a snow squall, and during subsequent movements the raider found herself in the midst of a British fishing fleet, but passed unrecognised in the darkness. And now that she was approaching the British coast, and the scene of actual operations, the barometer again obliged by falling rapidly.

It was a wild night and very dark when the first mine splashed overboard. A snowstorm set in, and as the work proceeded heavy seas broke over the vessel, smothering her with spray, but she was comparatively a large ship, built for ocean trade. Although the darkness and the snow were conditions favourable to the laying of mines in secret, and without their aid the danger of discovery would have been great, the rising gale and the heavy seas rendered the work both difficult and dangerous, notwithstanding that these deadly weapons were so arranged as to go automatically overboard.

Before the last of her cargo had been consigned to the deep it was blowing great guns, and one sea after another was breaking over the ship. Although sheltered waters lay less than fifty miles distant, to proceed there would mean certain discovery and destruction, so all through that wild night, and for many hours afterwards, the raider sought by every means in her power to battle seawards, away from the coast and danger, heading into the teeth of the gale and out on to the broad bosom of the North Atlantic, all unknowing that but for the severity of the storm she must have been observed, probably in the very act of laying the mine-field, by the small warship riding out the north-wester in the more sheltered waters close inshore.

****

It is interesting to note that it was on this mine-field a few days later that one of the largest transatlantic liners was sunk.

CHAPTER 21

The S.O.S.

A Great work of rescue was carried on throughout the war on all the seven seas by vessels of both the old and the new navy. This service was rendered to ally, neutral and enemy alike, but no complete record of the gallant deeds performed nor even of the numbers and nationalities of those saved will, in all probability, ever be available, and none is needed, for it was a duty which brought its own reward.

Typical of the way succour was brought by the naval patrols to those unhappy victims of both sexes left adrift in open boats in calm and rough, sunshine and snow, all over the northern seas by the cowardly *Unterseeboten* of the *kultured* race was the rescue of the passengers and crew of a liner off the wild west coast of Ireland in the winter of 1916.

****

It was mid-December, and flurries of snow were being driven before a stinging north-westerly wind. The sea was moderate, but the heavy Atlantic swell caused the lonely patrol ship to sink sluggishly into the watery hollows, with only her aerials showing above the surrounding slopes of grey-green sea, and a minute or so later to be poised giddily on the bosoms of acre-wide rollers with nothing but the white mists obscuring the broad horizon.

It was a wild wintry scene, pregnant with cold and hardship. The officer who had just come up from the warmth of the wardroom to relieve his "opposite number" on the bridge pulled the thick wool muffler closer round his neck and dug mittened hands deep into the pockets of his duffel coat.

In the Marconi cabin, situated on the deck of the sloop, a young operator was sitting with the receiving instrument fixed to his head and the clean and bright apparatus all around. He was city born and bred, and felt keenly the monotony of life at sea, although to him came the many interesting wireless signals from the vast network of patrols which covered the Western Ocean—linking the sea-divided units into a more or less homogeneous fleet.

Presently a message began to spell itself in Morse. Taking a pencil, the operator scribbled various hieroglyphics on the naval signal paper lying on the desk in front of him; then after a pause of a few seconds he pulled forward a tiny lever and began a rhythmic tap on an ebonite key.

It was the "S.O.S." call and the reply that had flashed through the ether. A minute or so later the written signal, giving the appeal for help and the position and name of the torpedoed liner, was handed to the commander. A glance at the chart told that young but experienced officer that he could not hope to bring his ship to the scene of the disaster before dusk closed down, and a message was sparked across the eighty miles of intervening sea asking how long the crippled ship could be kept afloat.

To this, however, there came no reply, and the engines of the sloop were put to full speed ahead. A heavy spray now commenced to sweep across the deck in drenching showers, and the snow haze thickened. The pitching of the ship increased as she raced over the ocean swell, driving her sharp bows deep into the masses of sea. The limbs of the watch grew stiff and numb, and a fine coating of wet salt stung their faces. Eyes ached from gazing into the bitter wind, and for over four hours the race against approaching night continued. If darkness closed down before that eighty miles of sea was covered all on board realised that the chances of finding any survivors would be greatly diminished. Even the strongest vitality could not long resist exposure to the intense cold, and there might be women and children in the sea ahead.

Many of the officers and crew of the sloop had experienced the agonies of cold, wounds and salt water when cast adrift on wintry seas, and the memory acted like a whip. As the hours went by the greenish tint of the sea slowly turned to leaden-grey, and the pure white of the driving snow contrasted sharply with the quickening dusk of the December night.

It was in the last half-hour of the dog watch that the sloop reached the scene of the disaster and the speed was reduced. Scattered over the sea around, and floating southwards in grim procession, was a mass of wreckage—a broken raft, a number of deck-chairs, spars and cordage, a life-belt and some oars—but of boats with living freights there was not a sign.

Steaming slowly round in widening circles, the sloop searched while the light lasted, but the whirling haze of fine snow blotted out the distance, and soon the early darkness of a winter night settled over the sea. The cold became intense. The white beam of a searchlight now flashed out over the black waters. There was a grave risk in this betraying light, one not sanctioned by the theory of war. It made the warship a target for any hostile submarine lurking around, but it seemed impossible to believe that a 6000-ton liner, with probably several hundred human beings on board, could have been so completely obliterated, and to the commander of the sloop the risk seemed justified.

Other ships might have intercepted the S.O.S. call and reached the scene of the disaster earlier, but the sloop's wireless, although put into action, could not confirm this, and so the search was continued.

On and off during the bitter night the white beam of light flashed out through the snow. For a few seconds it swept the sea close around and was then shut off. In the pall-like blackness which followed ears listened intently, but could distinguish nothing except the lash of the sea.

The sound-deadening qualities of falling snow would have cut short the range of any cry, for the human voice at its strongest, and with the atmospheric conditions favourable, can seldom be heard more than 1000 yards distant. So hour after hour of numbing cold went by with nothing to show except the occasional pathway of light on the grey slopes of sea and the low moaning wind.

The snowing ceased, and in the cold stillness which so often precedes daybreak in the north a faint cry came from the sea, at first so indistinct and mingled with water noises that it would never have been heard at all if the engines of the sloop had not been shut off, as they had been at frequent intervals during the night, to enable those on board to listen. The cry was quickly followed by the "snore" of a boat's fog-horn. A few turns of the sloop's propellers and in the grey light of the December dawn a large ship's life-boat could be dimly seen, away to starboard, when it rose on the bosom of the swell.

Careful manoeuvring placed the warship alongside the boat-load of half-frozen castaways and the work of rescue commenced. It was a sad task. Amongst the thirty-two survivors there were twelve women and children, seven of whom had died of cold and exposure during that bitter night. One, a young Canadian wife coming home to her wounded soldier husband, had been crushed by the explosion of the first torpedo and suffered agonies in the open boat before sinking into the peace of death.

To dwell here on the suffering caused by intense cold, exposure, hunger, thirst, untended wounds, and the mental agony of suspense, often to delicate women and children, when cast adrift on the open sea, would be merely to repeat what has so often been written, and which will live for ever in the memory of sailormen.

When the survivors had all been lifted on board—and many had suffered badly from frost-bite—the search for two other life-boats which it was learned had succeeded in getting away from the wrecked liner was commenced.

Shortly before midday the snowing began again and the wind moaned dismally through the rigging. Spurts of icy spray shot upwards from the bows and were blown back across the fore-deck of the ship, searing the skin of the tired men on watch. For several hours the sea around was searched in vain. Flurries of snow obscured everything more than a few hundred yards distant. Then towards four bells the storm passed and the air cleared of its white fog, but nothing was visible except the wide sweep of colourless heaving sea and leaden sky.

It came suddenly—an indescribable explosion with a violent uprush of water, followed by the hoarse shouting of orders, the low groans of wounded men and the sharp crack of cordite. The bows of the sloop had been blown off by a torpedo, and the vessel commenced to rapidly settle down.

The two undamaged boats were lowered and the survivors from the liner once again cast adrift to face the horrors of the previous night. Rafts floated free with all that were left of the crew of the sloop—two officers and thirty men. Their condition was pitiable. There had been no time to get either food or extra clothing, and so heavily laden were the light structures of *capuc* and wood that the occupants were continually awash.

Barely had the boats and rafts got clear of the ship before she took the final plunge, going down in a cloud of steam. A few minutes later the U-boat rose to the surface about 300 yards distant, and after remaining there for some time, without making any effort to render assistance, she steamed slowly away.

The boats took the rafts in tow, and the wounded, who suffered terribly from the cold and the salt water, were all transferred to the former. One of the women survivors from the torpedoed liner collapsed during the first hour, and although given extra clothing cheerfully discarded by the men, she died soon afterwards.

Seas washed over the rafts and sent clouds of stinging spray into the crowded life-boats. A biting frost stiffened the wet garments, which rasped the raw and bleeding wrists of the men who tugged at the oars—partly to increase their circulation and partly to keep the boats head-on to the sea. The only hope of rescue lay in keeping afloat until daylight, when the "S.O.S." call sent out before the sloop foundered might bring them aid. The coast of Ireland lay 300 miles to the south-east, and so intense was the cold that few expected to live through the night.

The gloom of a winter afternoon gave place to darkness, and with the fading of daylight the cold increased. Men became numb and were washed unnoticed from the rafts. Others were dragged unconscious into the already overcrowded life-boats, which sank so deep in the water with the additional weight that green seas now splashed inboard and baling became necessary. Limbs stiffened in the sharp frost and had to be pounded back to life by unselfish comrades. Even under cover of the sails the cold was so intense that only five women and two children were left alive by midnight.

Through the long dark hours men struggled under the drenching showers of bitter spray. When dawn broke, throwing a pale mystic light over the acre-wide Atlantic swell, each one knew that life depended on the coming of a ship before the light of day again faded in the west.

The snowing had ceased some hours before darkness lifted, and in the clear morning cold men stood up painfully and searched the watery horizon for the sign which would bring them life. Just before three bells, as the boats rose on the bosom of the swell, a thin blur of smoke could be seen low down on the eastern horizon. Had there been strength left in the worn-out bodies there would have been a cheer, but now only a slight stir of suppressed excitement and many a silent prayer.

The limit of human suffering and endurance had, however, not yet been reached. Some twenty minutes later it became evident that the ship had not received the wireless call and was passing too far off to be reached by any sound signal short of a big gun. Slowly the trail of smoke disappeared in the haze of great distance without even a glimpse of the ship itself.

The spirits of all began to sink as hour after hour went by without sight of the hoped-for sail. Then, about eight bells, one of the men standing up in the centre of the first officer's boat gave a little inarticulate cry and some few minutes later the dim outline of a big ship hove in sight. The suspense was unbearable. Women to whom any sign of religious emotion was alien knelt openly and prayed, while men who had suffered similarly before gazed fixedly at the distant object, knowing how fickle is Fortune to sailormen in distress. But the hull grew larger and hope shone on the faces of all. Men pulled frantically at the oars, while others waved pieces of sail or clothing to attract attention.

Now came a surprise. From the pocket of his duffel coat the first officer produced what he had hitherto kept hidden for just such an emergency—a Very's pistol, with its small-sized single red rocket. A hoarse cry of joy went up from all in spite of their exhaustion when they saw the rocket soar into the air and burst into a blood-red glow.

A short time later keen eyes made out the string of flags which fluttered from the halyards of the oncoming warship, and although minutes seemed like hours, none could quite remember what happened after. Some say that the cruiser came alongside them and others that she lowered her boats and steamed round in a circle. But forty-eight survivors were landed in Liverpool three days later, leaving in the wastes of the Western Ocean a murdered two hundred.

****

It is interesting to note that survivors from torpedoed ships frequently showed great reluctance to leave their life-boats and go aboard the rescuing vessel, especially when they were within easy sailing distance of a harbour. After being torpedoed, rescued and torpedoed again they often preferred the comparative safety but hardship of the small open boat to the risk and luxury of the big ship. This applied more especially to Scandinavian sailors, whose powers in small boats are well known.

It should, however, be stated that, so far as British and American seamen were concerned, men sailed again and again, after being torpedoed or mined six, seven and even eight times. It was this remarkable fortitude of the Mercantile Marine which saved Europe from starvation.

CHAPTER 22

In the Shadow of a Big Sea Fight

On the evening of 30th May 1916 six of his Majesty's drifters were lying alongside the quay of a Scottish naval base having their few hours' "stand-off" after weary days patrolling lines of submerged nets. Their officers and crews, with the exception of one sad-faced company on guard duty, were enjoying either the comparative luxury of a corrugated-iron wardroom, situated on a windy stone pier, or a few the more complete relaxation of a brief visit to a theatre in a neighbouring town. There were also many other ships coaling, resting and being repaired, for the base was a large and important one.

In the intelligence office an assistant paymaster, weary of decoding cipher wireless messages from flotillas, patrols and sweepers spread far out over the leagues of sea lying between this port and the German coast, sat talking to the executive officer on night duty.

About 8 p.m. a messenger from the wireless cabin entered with the familiar signal form and the A.P. spread it out carelessly on the desk in front of him, taking the sturdy little lead-covered decipher book from the safe at his side. A few scratches of the pen beneath the secret signal and the deciphering was complete. He looked up quickly and with a gesture of keen satisfaction handed the signal to the officer temporarily in command of the base.

The older man read it and paused for a moment before replying. It was the brief and now historic statement that an action between Sir David Beatty's battle cruisers and the German High Seas Fleet was imminent. A crowd of orders to be executed in the event of all kinds of emergencies were rapidly reviewed in his active brain. For a brief space the scene of what was occurring out in the blackness of the North Sea occupied his thoughts, for he had fought in the battle of the Dogger Bank and knew what those brief words really meant. It was the evening of the battle of Jutland.

Rising quickly to his feet, the night duty officer seized the telephone, rang up the Admiral Commanding, who had gone home to dinner, and hurriedly left the intelligence office to carry out a host of prearranged orders.

The "old man," as admirals are invariably called, was evidently ready for the emergency, for his large grey car tore past the sentries at the approaches to the base, and in a few minutes he was closeted with his commanders and other officers in the small matchboarded cabin. Charts were pinned down on the table in front of him, and for the next half-hour officers and messengers were kept busy with telephones and other means of rapid concentration.

In the neighbouring large town the police had received the order for a "general naval recall" and were active in the streets politely informing officers and men on short leave that their services were required immediately at the bases. In the theatres and cinema halls the cryptic message, "All naval officers and men to return at once to their ships," was given out from the stage or thrown on the screen, a replica of the night before Waterloo.

Men wondered and women grew anxious. Did it mean an invasion or an air raid? Many were the questions asked as silently seats were left and files of blue and gold streamed out of the places of amusement. Taxi-cabs full of officers raced each other along the streets. Civilians had to give place to sailors on the tram-cars, and then, in less than thirty minutes, all was quiet again, except for groups of people discussing possibilities in front of the big public buildings. Even these soon dispersed when reassuring messages were circulated which hinted at the reason for the recall, and the level-headed Scottish citizens went home wondering what the great news would be on the morrow—for the fate of empires might be decided during the night.

As each officer and man entered the base the gates were closed. The sentries and the officer of the guard knew nothing "officially," but in the wardroom at the end of the stone quay the news of the action was being discussed in imaginative detail. At 11 p.m. orders were received for certain small ships to get under way with sealed orders. An hour later came the message that six drifters were to be cleared of all their war appliances and were to be given stretchers, cots, slings and other appliances for the carriage of wounded. They were to be ready to proceed to sea at 2 a.m.

All was ordered hurry. Piles of anti-submarine devices were taken from the holds of these ships. Other vessels came alongside and unloaded stretchers, cots and slings, which had been obtained from local naval hospitals and hospital ships. The officers were grouped round a commander in the wardroom having typed orders, which had evidently been prepared long beforehand, carefully explained to them. Red Cross flags were served out, and by 1.30 a.m. all were ready for sea.

Other ships stole silently out into the blueness of the night to strengthen patrols and prevent hostile submarines from getting into position to attack the main battle fleets on their return to harbour.

Wireless messages indicating a concentration of German submarines on the lines of communication were received. Every armed ship was in great demand, but over the dark waters, lashed by a stiff easterly breeze, the gunners of many batteries gazed steadily as the searchlights played around, investigating everything that moved on the face of the waters. Beams flashed heavenwards for hostile aerial fleets.

On the dark quaysides and on the decks of the ships hundreds of sailors moved noiselessly about getting ready for sea. Columns of smoke from the short funnels of destroyers, trawlers and drifters showed up black against the indigo void, and ever and anon hoarse voices shouted orders, unintelligible from the distance. It was quiet preparation rather than noisy haste, and although an air of suppressed excitement did prevail when the men were mustered and extra hands told off to the different ships by the light of battle lanterns, it was more a feeling of hope than one of satisfaction.

For nearly two years these men had quietly fought the elusive submarine, the nerve-shattering mine, and endured uncomplainingly the terrible hardships, arduous work and monotony of patrol, and now their one fervent hope was a glimpse at least of the real thing.

In the wardroom on the quay about sixty officers of all ranks were discussing the possibilities of the fight while waiting impatiently for the last command before the relief of action—"Carry on as ordered." Conversation centred on the Grand Fleet, under Sir John Jellicoe, steaming down from the north. Many had seen those miles of gigantic warships, whose mere existence had preserved for the Entente the command of the sea and all that it implied. Others had served in ships whose names have been familiar to Englishmen since the days of Nelson, and now opined that when at last the "old ship"—perhaps a brand-new super-dreadnought—was going into action on the great day it was their luck to be in command of a "one-horse" boat miles from the field of glory.

Four bells had struck when the signal came for all ships under orders to proceed to sea. Oilskins were rapidly slipped on, for a fine rain had commenced to fall and the damp wind was penetratingly cold at this early hour. Almost silently the small grey ships slid out of harbour and were lost in the blueness of the night.

****

When dawn broke over the choppy tumbling sea the different flotillas were far apart, each attending to its allotted task. Those engaged in patrolling the route by which the battle cruisers would return found themselves acting in conjunction with a division of destroyers, some of whom had been under refit but a few hours previously, but when the tocsin of battle rang out had *made themselves* ready for sea in an incredibly short time, thereby earning the praise of the commander-in-chief.

Information had been received, and later in the day was confirmed, that no less than five hostile submarines were known to be waiting in the vicinity with the object of attacking any crippled ships from the battle fleets, and it became the duty of the patrols to clear them away from the lines of communication. For over twenty hours the seas around were churned by the keels of a heterogeneous fleet of ships armed with equally heterogeneous weapons. Guns' crews stayed by their weapons until their limbs ached and look-outs searched the sea with burning eyes. Through the short dark hours of a May night in northern latitudes searchlights swept the near approaches, while in the black void of sea and sky beyond myriads of mosquito craft moved over the face of the waters with all lights out and their narrow decks cleared for action. Alarms were frequent, and the occasional yellow flashes and sharp reports of cordite, some too far distant to be visible, told their own tale. In the treacherous light of early dawn the fins of big porpoises were more than once mistaken for the hunted periscope.

****

With the Red Cross flotilla waiting behind the screen of patrols and defences things had moved rapidly. Each little ship had been told off to attend on one or other of the great warships which were hourly expected from the battle zone. Stretchers, bedding, cots and slings were piled on the decks, and extra hands had been lent for the work of removing the wounded.

Another flotilla was in readiness to replace the casualties with reinforcements, which had been concentrated by special trains, in order that the battle fleets and squadrons might be again ready for sea in the shortest possible time.

At the base trains and big ships were waiting with every known appliance to alleviate the suffering which was coming in from the sea.

It was a typical May morning, with a light easterly breeze, when the first of the great line of ships—H.M.S. *Lion*—came into view. A hurricane of cheers greeted her from the deck of every ship that passed. Then the gallant *Warspite*, low by the stern and scarred and torn by tornadoes of shell; the *New Zealand*, scarcely touched by the fiery ordeal; the plucky little light cruiser *Southampton*, holed and battered; followed by cruiser after cruiser with attendant destroyers, some with great bright steel splinters of shell still sticking tight in the gouged armour-plate; others with holes plugged with wood and broadsides stained with the bright yellow of high explosives. Gun shields caught by the gusts of shell were cut out like fretwork; funnels were blotched with blackened holes; but of them all not one was out of action. Few, if any, of the heavy guns and armoured barbettes were damaged, and all except one—the *Warspite*—came in proudly under their own steam. This was the return of the battle cruiser and light cruiser squadrons, which, under Sir David Beatty, had met and defeated practically the entire German navy. Steaming back into the northern mist was the Grand Fleet—the largest assembly of warships ever known—which, had it been given the opportunity so eagerly sought, would undoubtedly have annihilated the remains of Von Hipper's fleet.

An observer from a distance would have found it difficult to believe that this was the fleet which had just fought the greatest sea fight in the history of the world. Yet the decks of the seaplane carrier *Engadine* were covered with men in motley clothes, a grim reminder of the severity of the ordeal, for they were the survivors from the thousands who had manned the *Princess Royal* and *Invincible*. On the high poop a fleet chaplain was surrounded by figures in borrowed duffel suits giving thanks to the God of Battles for their rescue.

As the engines of each great ship came temporarily to rest a vessel of the Red Cross flotilla ranged alongside and the more sombre work of war began. A shell through the sick-bay of H.M.S. *Lion* had caused Sir David Beatty to have many of the wounded on that ship placed in his own cabins. The only casualty on the *New Zealand* was caused by a gust of bursting steel over the signal bridge. A big shell had passed longitudinally through the line of officers' cabins in the battered little *Southampton*, and many were the curious escapes from death. In modern naval war a heavy casualty list seems unavoidable, and the deadly nature of a sea fight will perhaps be better realised when it is stated that on one of the battle cruisers there were just over three hundred casualties, of which number very nearly two hundred were killed outright, and this on a ship which still sailed proudly into port in fighting condition. Where the shells had burst in the steel flats the fierce heat generated had burnt off the clothes and skin of many who were untouched by the flying slivers of steel, and the crews of the secondary batteries of smaller guns suffered severely.

Cot cases were the first to be lowered from the decks of the warships to the waiting Red Cross boats. The patience and care with which this difficult operation was carried out may be gauged from the fact that there were no casualties or deaths during the work of transportation. Human forms, swathed from head to foot in yellow *picric*-acid dressings, were lowered on to the decks or carried down the gangways. By a curious ordinance of fate, *picric acid*, one of the most deadly explosives known, also provides a medical dressing for the alleviation of the pain which in another form it may have caused. The walking wounded, with arms in slings or heads covered in lint, were helped down the ship's sides by smoke-blackened comrades in uniforms torn to shreds by the fierce work of naval war.

All serious cases of shell shock were conveyed at the utmost speed by special units to the big and lavishly equipped hospital ships. Those with minor injuries were taken ashore and placed in ambulance trains for distribution among the big naval hospitals. So perfect was the organisation that within three hours all the sick-bays had been cleared and fresh crews placed on board. The squadrons were again ready to give battle.

Twenty-four hours later the patrol flotillas returned to their base to commence once again the dangerous and monotonous but less spectacular work of minesweeping and patrol. Their work in preventing a concentration of German submarines on the line of route of the returning fleets and in the removal of the wounded received high praise from the commander-in-chief. In the wardroom on the little stone pier a silent toast was given that night to those who had gone aloft in the greatest sea fight since Trafalgar.

CHAPTER 23

A Night Attack

Two drifters, about a mile apart, with no lights to indicate their presence, were drifting idly with the ebb tide. It was an oppressively hot night in mid-August. Scarcely a ripple disturbed the surface of the sea, but the intense darkness and the absence of stars told of the heavy clouds above. The barometer had been falling rapidly for some hours and all the conditions seemed to indicate a coming storm.

The duty of these two vessels was to watch lines of cunningly laid submerged nets (described in an earlier chapter) and to guide the few merchant ships which passed that way through the labyrinth of these defences, laid temporarily as a trap for the wily "Fritz" if he should chance to be cruising in the vicinity.

The drifters were adequately armed with guns and depth charges to attack any such monster of the deep which betrayed its presence by becoming entangled in the fine wire mesh and so attaching to itself a flaming tail, which would then be dragged along the surface, marking it as a target for all the pleasant surprises lying ready on the decks of the patrols.

Fishing for Fritz was a popular sport in the anti-submarine service until the "fish" became shy and its devotees *blasé*; then the primitive net was changed for the more scientific devices already described. It required infinite patience and meant very hard work, with a *soupçon* of danger thrown in. For when the tons of steel wire-netting, with its heavy sinkers and floats, had been laid, days were spent in watching and repairing, then endless resource employed to induce a submarine to enter the trap.

Occasionally the voyage ended in an exciting chase, with the flaming buoy as the guiding light.

It was in the early period of the war, when Paris was still threatened by the Teutonic armies and the Allies waited confidently for the clash of the great battle fleets. Every dark night on the northern sea eyes and ears were silently watching and listening for the comings and goings which would herald the storm. The strain was great though the work was not spectacular, for all knew that the safety of England, or at least its freedom from invasion, might, for one brief historical instant, depend on the vigilance and nerve of that heterogeneous, irregular horse, the sea patrols.

The great cruiser squadrons were scouring the North Sea. Battle seemed imminent, and that vague wave of human electricity which passes along the firing line before the attack at dawn, and even extends to the lines of communication, was in the air on this dark night in 1915.

Six bells had just struck when a faint, cool breeze swept across the surface, and a few minutes later the first vivid flash of lightning forked the eastern sky. There was a scramble for oilskins on Drifter 42 as the rain came hissing down like a flood released. The storm was severe while it lasted. The thunder rolled over the placid surface. Lightning darted athwart the sky, illuminating the black void beneath. For about thirty minutes the sky blazed and roared, then the hiss of the rain ceased and the storm moved slowly northwards, but one of the final flashes revealed something low down on the surface moving stealthily forward. So brief was the glimpse obtained, however, that it seemed merely a phantom—by no means uncommon occurrences when men have been watching for years. When the next flash came the surface of the sea around was clear.

As was usual in such cases, half the watch on deck could swear they had seen it, while those who were not looking ridiculed the idea, so the C.O. said nothing and took precautions. The watch below was called and the powerful little gun on the fore-deck manned. Then all waited in silence, listening intently for the curious, creaking noise of a submarine under way.

In those early days of hostilities there were no elaborate hydrophones for detecting the approach of submarines under the water, and the only hope of a warning came from the possibility of the under-water vessel breaking surface momentarily. The uselessness of the periscope for navigation during darkness, which at present forms the principal limitation of submarines, made it distinctly likely that she would cruise on the surface at night, and if forced to dive would be more or less compelled to quickly rise again in order to ascertain the position of her enemy before it would be possible to fire a torpedo with any chance of success.

For these reasons all eyes and ears on the drifter were strained to catch the first glimpse or sound, and dead silence was maintained. It is in times like this that one discovers how acute the senses become when danger lurks in the darkness around. Things undetectable under normal conditions can be seen or heard distinctly when life depends on the intelligence so gained.

Long minutes of silence slipped by and nothing occurred; then came the distant and familiar creaking noise, almost inaudible at first. The gun's crew braced themselves for the stern work ahead. On the rapidity and accuracy of their fire not only their own lives, but also those of their comrades, would probably depend. The gun-layer bent his back and glanced along the grey tube to the tiny blue glow of the electric night sight. The shell was placed in the open breech. Then came those interminable seconds before an action begins.

The tension would have been almost unsupportable had nearly all of the crew not grown accustomed to life hanging in the balance on the wastes of sea.

A flicker of light, at first almost spectral, appeared from out of the darkness some 500 yards to starboard. It grew almost instantly into a bright white flare, illuminating the surface of the black water as it moved along. The pungent smell of burning calcium floated over the sea and the drifter's engines began to throb heavily.

The tension relaxed, a subdued cheer broke from the crew of the drifter as she gathered speed, and the Morse lamp winked its order for concerted action to the other drifter somewhere in the darkness around. An answering dot-dash-dot of light appeared from away to starboard and the chase commenced in earnest.

A few minutes later the glare from the calcium buoy, now being towed through the water at several knots, shone on the faces of the crew as they trained their gun ahead, but the submarine was under the surface and, although probably quite unaware of the flaming tail which was betraying her movements, appeared to know that she was being hunted by surface craft. After running straight ahead for a few minutes she turned eight points to the eastward in an attempt to baffle pursuit.

The chase was a fairly long one, as the speed of the drifters was not sufficient to enable them to gain rapidly on their quarry, but the flexibility of the steam-engine gradually gave the surface ships the advantage and they crept up level with the light. Then, with their boilers almost bursting and flames spouting from the funnels, they drew ahead until over the submarine itself. Depth charges were dropped from the stern of the drifters. The water boiled with the force of the explosions and the light on the buoy went out. Still the drifters held their course in the now pall-like blackness, and other bombs splashed into the water astern, to explode with a dull vibration a few seconds after they had sunk from the surface.

The engines of the two small surface ships were shut off and every ear became alert, but no sound broke the stillness of the summer night, except the rumble of distant thunder and the gentle lap of the sea against the sides. Morse signals winked from one ship to the other and back again. When due precautions had been taken against a further surprise attack, the chivalry of the sea called for a search to be made for possible survivors. This was done with the aid of flares, but only oil and some small debris were found. Dan-buoys were dropped to mark the spot and soundings taken. Twenty-four fathoms deep was added to the report of the action, and a few days later a diver reported having found the wreck of the U-C 00.

CHAPTER 24

Mysteries of the Great Sea Wastes

The piratical warfare of German submarines produced many sea mysteries. Some were solved after the lapse of months and even years, while others will, in all probability, remain unknown until the sea gives up its dead. Among the latter may be numbered the curious discovery in the North Atlantic of a nameless sailing ship, without cargo, identifying papers or crew, but sound from truck to kelson, and with her two life-boats stowed neatly inboard and a half-finished meal on the cabin table. Experts examined this vessel when brought into port, but so far have been utterly unable to offer any solution or discover any clue, beyond the fact that she was built and fitted out in some American port and carried an unusually large crew.

Another similar mystery was the disappearance of a French vessel while on a voyage to New Orleans and the discovery eleven months afterwards that she had called for water and food at a small port on the Pacific coast of South America. No further trace has so far come to light, nor the reason for her changing course and rounding Cape Horn. A mystery which remains a mystery to the end of the chapter is likely to be irritating to the imaginative mind, but to the following occurrence there came a solution after the lapse of a few weeks.

THE SPECTRE OF THE GOODWINS

It was a pitch-black night, with fine rain driving up from the south-west. The summer gale which had raged for the past twenty-four hours had blown itself out, and although the steep seas still retained their night-caps, the wind came only in fit-ful gusts. Away to starboard an indistinct blur of white foam stretched athwart the sea and the dull roar from the maelstrom of the Goodwins rolled across the miles of intervening water.

The armed trawler *Curlew* bravely shouldered her way through each green comber as it rose to meet her, lurching over the seas in a smother of spray. Oilskinned figures moved warily along the life-lines, for when a wave struck her tons of water swept across her slanting decks, submerging the bulwarks and causing the sturdy ship to groan and tremble from stem to stern.

In the little bridge-house the dim light from the binnacle shone on the hard wet face of the commanding officer, who watched the seas as they rose up ahead, giving directions to the man at the wheel, and all the while keeping a watchful eye on the distant blur of foam covering the treacherous shoals.

Few except sailormen can realise the dangers and anxieties of navigation in times of war. The absence not only of the warning lights which in days of peace flash their signals far out over the seas, marking the innumerable dangers which lie along treacher-ous coasts, but also of warships and merchantmen rushing through the night with not even the flicker from a port-hole to denote their coming—perhaps at a speed of nearly three-quarters of a mile a minute; a second's indecision on the part of the brain and nerve directing each ship, a momentary forgetfulness of that elu-sive "right thing to do"—some second danger to attract a flash of attention from the first—even a blinding cloud of spray at the psychological moment and, well, two more ships have gone, with perhaps hundreds of lives. Yet these things but seldom happened, and the reason was that all that tireless energy, skill and nerve could do was done on the sea in those years of storm and stress.

Some two hours later, and just before dawn broke over the tumbling sea, an exceptionally heavy wave struck the trawler full on the port-bow. The hammer-like blows of the water as it poured on board and struck the base of the wheel-house and superstructure momentarily drowned all other sound. When the air had cleared of flying spume a big black hull loomed out of the darkness ahead and seemed suddenly to grow to an immense size, towering high above the trawler's forecastle-head. A blast on the whistle, a sharp order and the trawler swung off to starboard, with the great black mass perilously near. It was a close shave, and the watch held their breath while waiting for the crash and shock which for a brief second seemed inevitable.

There was no time for action or signal. The great ship slid past like some black phantom framed in the white of flying scud. It faded into the misty darkness of sea and sky almost as quickly as it had appeared, and, curiously, no sound of throbbing engines accompanied its passage. It took the captain of the patrol but a minute to make up his mind what to do. He gave a quick order to the helmsman and a warning shout to the watch below on deck. The little ship, as she came about, lurched into the trough of a sea and rose shivering from end to end. The next moment an avalanche of white and green water poured over her, flooding the decks and sending clouds of spray high over the funnel and masts. Then commenced an exciting chase, with the seas racing up astern and all eyes trying to penetrate the darkness ahead.

The faint misty light of a new day had brightened the eastern horizon before the mysterious ship again loomed up ahead. The heavy sea still running made it difficult, however, to distinguish any national or local characteristics which might give a clue to her identity or intentions, and the suspense was keen.

The two guns of the patrol vessel were manned, and a three-flag signal fluttered from the jumper-stay but received no immediate reply from the ship ahead. Then, after a few minutes' pause, during which time the trawler manoeuvred for the advantage of the light from the breaking dawn, a yellow flash belched from her side and a shell ricocheted off the water just ahead of the mysterious steamer. Still there was no response; but it could now be plainly seen that the engines were not working and that she was drifting before the wind and sea.

Was it merely a *ruse de guerre* to gain the advantage in the event of an attack, or was she a vessel disabled by the storm which had raged during the past forty-eight hours? Neither of these suppositions, however, satisfactorily explained the total disregard of signals and the warning shot which had been fired across her bows.

Again a line of flags were hoisted on the trawler's halyards, this time a well-known signal from the *International Code*, but still no notice was taken of the peremptory order it conveyed.

After the chase had been on for over an hour another shot was fired from the trawler. The report echoed across the still boisterous sea and the splash of the shell just cleared the ship's bow. Still there was no response, and the trawler's course was altered so that she would soon close in on her quarry. As the light increased it was seen that a stout wire hawser was trailing in the water from the starboard bow, and suspicion of some new evidence of sea *kultur* increased. When the range had closed to about 1000 yards she slowly swung round until almost broad-side-on to the trawler, whose guns instantly opened fire in earnest. The third shell struck the large wheel-house of the mystery ship, demolishing it completely. When it became evident that the fire was not going to be returned, the guns of the trawler again ceased, and the two vessels drew close to each other. A partly defaced name, which was rendered indecipherable by the splash of the seas as they struck the counter, could be distinguished with the aid of binoculars in the quickening light of early morning, but neither officers nor crew could be seen, the bridge and decks appearing deserted.

Not to be misled by this ruse, however—for on similar occasions ships had been blown to pieces at close range by concealed batteries—the *Curlew* approached cautiously, bows-on, offering the smallest possible target, and with her guns trained on the quarry. This sea-stalking is nervy work and must be played slowly. Twice the trawler circled round the mysterious ship, and the sun had mounted high, penetrating the banks of cloud which scudded across the summer sky and tingeing the still boisterous sea with flecks of golden light, before it was considered safe to relax all precautions. Even then the sea prevented any attempt being made to board the curious craft, and for six hours the trawler clung to the heels of her quarry, which was rapidly drifting far out into the North Sea.

The danger of attack from hostile submarines was great, and the gunners stood by their weapons although drenched every few seconds by the floods of heavy spray which still poured over the bows. At last patience and endurance were rewarded. The sea calmed sufficiently to enable a boat to be lowered and with difficulty brought up under the lee of the mysterious ship.

An armed guard, headed by the sub-lieutenant, eagerly scrambled up the lofty rolling sides. They had scarcely reached the deck before their only means of retreat was cut off. The two men left in the life-boat were unable to keep her off the iron sides of the big ship. She rose like a cork on the crest of a wave until almost level with the top line of port-holes and then dropped back, catching the edges of the rolling-stocks. There was a crash of splintering wood and the next minute two half-drowned men were being hauled up the sides by their comrades on deck.

It was an anxious moment, for although the decks seemed deserted there was that curious, uncanny feeling which is ever present when facing an unknown peril. After all it *might* prove to be a *ruse de guerre* or some new form of frightfulness. There were only six men from the trawler—a small enough party, however well armed, if it came to a fight—and great caution was observed while exploring the ship. A signal had been arranged in the event of treachery, and the *Curlew*, with her guns *and wireless*, would prove a dangerous antagonist.

All was well, however, for the ship was deserted. A careful inspection of the cabins showed that the departure of officers and crew had been a hasty one, but all the ship's papers had been carefully removed. The forepeak or bow water-tight compartment was full of water, but the bulk-head had held and kept the vessel afloat. Beyond this no damage was visible above the water-line and the condition of both hull and engines was good. She proved to be a Spanish ship, and to make the mystery deeper her four life-boats were still on the davits, although swung outboard ready for lowering.

In those troublous days the fact of the life-boats being hoisted out in readiness for eventualities conveyed little or nothing, but when a careful search proved that many of the life-belts had gone with the crew the problem became an interesting one.

Had they been taken on to the deck of a German submarine which had subsequently dived and left them to drown, as was the case with the crew of a British fishing vessel, or had they been conveyed as prisoners of war to Germany? Against both of these surmises was the fact that *all* the ship's boats remained, and a German submarine would scarcely be likely to come close alongside even a neutral ship, especially during the bad weather that had prevailed for the past few days. Would it remain one of the many mysteries of the great sea war?

Some few hours later the trawler, with her big "prize"—under her own steam—entered an eastern naval base and berthed her capture with the aid of tugs.

****

The explanation came from headquarters several weeks later. The S.S. —, of Barcelona, had grounded on the Goodwins about three hours before she nearly ran down the trawler. Her crew, thinking that she would rapidly break up in the surf, had fired distress signals and been taken safely ashore in a life-boat. The rising tide and south-westerly wind had done the rest, freeing her from the dangerous sands.

CHAPTER 25

From Out the Clouds and Under-Seas

It has already been shown that the science of aerial warfare is closely allied with that of under-sea fighting. Airships and seaplanes play important parts in all anti-submarine operations. They make very efficient patrols and can detect the presence of both submarines and mines under the surface.

During the Great War there were stations for armed aircraft all round the British coast, and the patrols of the sea and air acted in close co-operation. It often happened that one was able to render important service to the other. An occasion such as this took place off an east coast base in November, 1916.

SALVING AN AIRSHIP

A big car dashed up the wooden pier of a small seaport regardless of the violent jolting from the uneven planking. It was pulled up with a jerk when level with one of the little grey patrol boats known by the generic name of M.L.'s, which was lying in the calm water alongside with its air compressor pumping vigorously.

Two officers of the Royal Naval Air Service, with a P.O., carrying a powerful Morse signalling lamp, jumped from the car and scrambled down the wooden piles on to the deck of the M.L.

A nod from the commanding officer and the mooring ropes were cast off as the telegraph was jammed over to "half ahead." Instantly the powerful engines responded to the order and the little ship began rapidly to gather way. When the harbour bar had been crossed the order for full speed was given and the engines settled down to a low staccato roar as they drove the M.L. over the heaving swell.

No word had yet been spoken between the officers of the sea and air. A brief telephone message to the little hut on the quayside from the adjacent naval base to the effect that M.L.A6 was to be ready to embark two officers from the air station and was to proceed in search of an airship which was foundering about twenty miles seawards was all that had been told, and yet not a single second of time was lost in getting under way. All recognised that it was a race to save the lives of men.

The little ship cleft the seas, smothering herself with foam, and bluish fumes poured out of the engine-room ventilators. The first half-hour seemed interminably long, and the horizon was continually searched with the aid of powerful glasses for a sign of the wrecked airship. At last a faint speck became visible away to the south-west, and as the distance slowly lessened— terribly slowly, notwithstanding the speed of nearly half-a-mile a minute—the crumpled envelope settling on the water could be distinguished.

It was a question of minutes. Again the order was shouted down the speaking-tube for more speed, but this time there was no reply. The C.O. rang the telegraph viciously, but without result. The coxswain at the wheel looked up quickly and then shouted an order to a deck hand, who lowered himself down the tiny man-hole in the deck leading to the engine-room. A few seconds later the second engineer appeared at the top of the fo'c'sle hatch and, ducking to avoid a heavy shower of spray, scrambled aft and peered down the man-hole, from which blue fumes, somewhat thicker and more pungent than usual, were rising. The next instant he too disappeared below.

The air officers were trying to get into communication with the rapidly sinking airship by means of the powerful Morse lamp, but without result, and one of them put his head into the wheel-house and asked anxiously if more speed was possible.

Just then the second engineer and one of the crew crawled out of the man-hole, pulling a limp figure behind them. The

C.O. turned to ascertain what had happened, and the men, very white and shaky, explained in a few gasps that they had found the chief engineer senseless at the bottom of the iron ladder leading up to the deck, and had themselves been nearly gassed by the petrol fumes.

Glancing at the blue vapour now pouring up the hatchway and out of the ventilators, the C.O. realised the risk of fire and explosion he ran by carrying on at such high speed, but he also knew that men were drowning in the sea some eight miles ahead, and that the few extra knots might make the difference between life and death for them.

That the risk must be taken was a foregone conclusion, but how to keep the engines running at that high speed without attention—for it was evident that no man could live for many minutes in the poisonous fumes—was a more difficult problem. This was solved, however, by the second engineer volunteering to go below with a life-line attached, so that he could be hauled up to the deck when giddiness came on. More than once this gallant petty officer had to be pulled up choking and exhausted. He risked instant death from the explosion of the gas from the leaking and overheated pipes and engines, as well as suffocation from the fumes, but he stuck to his post, returning again and again into the poisonous atmosphere.

Darkness was gradually settling over the sea, and the flickering light of the Morse lamp—still asking for a reply—made yellow streaks on the wet fore-deck. Presently a faint speck of light blinked amid the dark mass of the airship, but almost instantly went out, and for some time nothing further was seen.

Barely three miles of heaving sea separated the two ships when the bright glare of a Very's light, fired from a pistol, soared into the air. A cheer broke from the dark figures on the deck of the M.L., and a message of hope was eagerly flashed back.

The last knot seemed a voyage in itself, but eventually the great dark mass of the still floating envelope loomed up ahead, and almost instantly the clang of the engine-room telegraph, shutting off the leaky engine, gave relief to the plucky second engineer, who had retained consciousness and control through that dreadful twenty minutes by frequently filling his aching lungs above the hatchway.

The sea around was a mass of tangled wires, in which the mast and rigging of the M.L. was the first to become entangled. Near approach was impossible, so orders were given to lower away the boat. The sturdy little steel-built life-boat splashed into the sea alongside, one minute rising on a wave high above the deck-line and the next disappearing into the dark void below. Figures slid down the miniature falls to man her and the next minute were pulling through the tangled wreckage to where the beam of the M.L.'s searchlight showed six airmen clinging to a floating but upturned cupola.

Numbed with the cold, they fell rather than jumped into the boat as it was pulled alongside. One was insensible and the others were too far gone to utter a word. Nothing but the wonderful vitality necessary to the airman as to the sailor had enabled them to hold on in that bitter cold for over two hours after eight hours in the air.

The task of extricating the M.L. from the tangle of wire stays and other wreckage was a difficult one. A propeller had entwined itself and become useless (afterwards freed by going astern), the little signal topmast and yard had been broken off by a loop of wire from the gigantic envelope and the ensign staff carried away. After about twenty minutes cutting and manoeuvring, however, she floated free, and a question was raised as to the possibility of salving the airship.

By this time another M.L., sent out to assist in the work of rescue, had arrived on the scene, and a conference between the air and sea officers on the senior ship resulted in the attempt at salving being made. Wires that were hanging from the nose of the airship were made fast to the stern of the M.L.'s, and all wreckage was, where possible, cut adrift. This, to the uninitiated, may sound a comparatively quick and simple operation, but when it is performed in the darkness, with the doubtful aid of two small searchlights, on a sea rising and falling under the influence of a heavy ground swell, it is anything but an easy or rapid operation, and occupied half the night.

The huge mass of the modern airship towered above the little patrol boats like some leviathan of the deep. To attempt its towage over twenty miles of sea seemed almost ludicrous for such small craft, and yet so light and easy of passage was this aerial monster that progress at the rate of three knots an hour was made when once the wreckage had been cut adrift, the weights released and the envelope had risen off the surface of the water.

Armed trawlers that passed in the night wondered if it was a captive zeppelin and winked out inquiries from their Morse lamps. A destroyer came out of the darkness to offer assistance. The cause of much anxiety had been the likelihood of hostile submarines being attracted to the scene by the helplessness of the airship, which had been visible, before darkness closed over, for many miles as she slowly settled down into the sea. This danger, however, passed away with the arrival of the destroyer and the armed trawlers, but another arose which threatened to wreck the whole venture. About 5 a.m. the wind began to freshen from the north-west and the M.L.'s towing the huge bag were immediately dragged to leeward. The combined power of their engines failed to head the airship into the wind and urgent signals for assistance were made to the destroyer and trawlers, who had, fortunately, constituted themselves a rear-guard.

A trawler came quickly to the rescue and got hold of an additional wire hanging down from the envelope. The destroyer, in the masterful way of these craft, proceeded to take charge of the operations. Her 9000-horse-power engines soon turned the airship into the path of safety, and with this big addition to the towing power it was less than half-an-hour later when the great envelope was safely landed on the quayside, much to the amazement of the townspeople.

"Unlucky Smith"

There is, however, another side to this co-operation between fleets of the sea and air. It has more than once occurred that vessels equipped almost exclusively for submarine hunting have been engaged by zeppelins, and actions between seaplanes and under-water craft have been frequent.

A MONITOR: THE BULGE OF THE "BLISTER" WILL BE SEEN ON THE WATER-LINE NEAR THE BOW.

How a large fleet of unarmed fishing vessels were saved and a zeppelin raid on the east coast of England prevented by the timely action of an armed auxiliary proves once again the truth of the old military axiom that it is the unexpected which always happens in war.

It had been one of the few really hot summer days granted by a grudging climate. The sea was a sheet of glass, the sky a cloudless blue, except where tinged with the golden glow of sunset. Lieutenant Smith smiled somewhat grimly as he mounted the little iron ladder and squeezed through the narrow doorway into the wheel-house. He nodded to the skipper—an old trawlerman acting as a chief warrant officer for navigational duties—as a signal for the mooring ropes to be cast off, and mechanically rang the engine-room telegraph. He had done all these things in the same way and at the same time of day for nearly two years. For a long while he had gone forth hopefully, saying to himself each cruise, "It's bound to come soon," but as the weeks grew into months, and the months promised to extend into years, disappointment gained the mastery and duty became appallingly monotonous and uninteresting.

This, however, did not cause him to work less strenuously or to neglect to watch the large fishing fleet which he guarded on four nights out of the seven, but each letter he received from old friends in other branches of the King's service brought tidings of excitement, rapid promotion, or at least a little of the pomp and circumstance of war, and he saw himself at the end of it all with nothing to show for years of danger, hardship and impaired health. The worry and the lonely monotony, trivial as he knew them to be, were slowly sapping his nerve and vitality.

The trawler glided from the harbour on to the broad expanse of tranquil sea, now aglow with the lights of a summer sunset. Slowly the coast-line faded into the blue haze of distance, and all around the watery plain was mottled with the shadowy patches made by the light evening breeze. Settling himself in an old deck-chair, which he kept in the wheel-house, Smith lit his pipe and allowed his thoughts to wander, but every now and then his eyes would search the sea from slowly darkening east to mellow west.

Although the summer was well advanced, there were but few hours of darkness out of the twenty-four in these northern latitudes, and when the armed trawler came in sight of the widely scattered fishing fleet, which it was her duty to guard throughout the night, a mystic half-light subdued all colours to a shadowy grey, but a pale amber afterglow still lingered in the sky and the stars were pale.

Smith lingered a few minutes on deck to finish a cigar before going below for his evening meal. Seldom during the past year had all the elements been so long at peace, and the contrast appealed to him as a luxury to be enjoyed at leisure. Even the light breeze of sunset had died away, leaving an unruffled calm, and the sails and stumpy funnels of the little fishing craft appeared like "painted ships on a painted ocean." For nearly an hour he sat inhaling the fragrant and satisfying smoke from more than one cigar, preferring the cool of the deck to the stuffy cabin. Then a dark blot appeared from out of the luminous blueness of the eastern sky and it travelled rapidly downwards towards his flock.

Smith watched it for several seconds, then it suddenly dawned upon him that the hand of the destroyer was coming even into this haven of peace, and a fierce resentment entered his soul. He heard the distant shouting of fishermen as they cut adrift their nets and prepared to scatter before the approaching zeppelin, and in a moment he realised that the long-awaited chance had come. It all seemed too unreal to be true, but he rose up quickly and in a few terse sentences gave the necessary orders for the guns' crews and engineers.

The whir of the airship's propellers grew rapidly louder and its bulk loomed black against the bright sky. Determined, however, to take no risk of failure, Lieutenant Smith withheld the fire of his guns until the great aerial monster, now travelling down to less than 1000 feet, was well within range.

Attracted by the helplessness of a large number of fishing craft congregated in a comparatively small area of sea, the destroyer dived to the attack like some giant bird of prey, unable in the gloom which shrouded the earth to distinguish the presence of an armed escort.

The suspense was painful. Then the muzzles of two high-angle guns rose up from the well-deck and superstructure of the armed patrol, and in response to a low-toned order from the C.O., giving the height, time and deflection, they quickly covered the great black body of their objective. Tongues of livid flame leapt from their mouths and were followed by sharp reports. A few minutes of heavy firing and the nose of the monster appeared to sag.

The men at the guns yelled exultantly, redoubling their efforts, and shell after shell went shrieking heavenwards. Suddenly the sea around rose up in huge cascades of foam and a shattering roar, which completely dwarfed the voice of the guns, shook the small ship from stem to stern. Everything movable was hurled across the deck. Breaking glass flew in all directions, and the aerials at the mast-heads snapped and came tumbling down with a mass of other gear. The cries of injured men arose from different parts of the ship, but still the guns hurled their shells, and the zeppelin, now well down by the head, rose high into the upper air and made off eastwards. After dropping all her bombs in close proximity to the armed trawler she had lightened herself sufficiently to rise out of range, but whether or not she would be able to keep up sufficiently long to reach her base, over 300 miles distant, was extremely doubtful.

Flames spurted from the short funnel of the patrol as she steamed at full speed after the retreating zeppelin, endeavouring to keep her within range as long as possible. It was a question of seconds. Before she finally disappeared in the increasing darkness another long-range hit was observed and the zeppelin receded from view, drifting helplessly.

The disappointment at not being able to give the *coup de grâce* to the aerial destroyer was keenly felt by all on board, for a half success is of little account in the navy. The gunners had done magnificently, the ship had been manoeuvred correctly and four of the crew had been wounded by fragments from the bombs dropped *en masse*, but notwithstanding their exertions and the luck which had brought the zeppelin down from the security of the skies, they had failed to secure the prize legitimately theirs.

That the attack on the fishing fleet had been successfully beaten off appeared a minor detail, and the voyage back to port in the quickening light of a beautiful summer morning was a sad pilgrimage. Scarcely a word unnecessary for the working of the ship was spoken, except Lieutenant Smith's brief explanation that it was just his luck.

****

About two weeks later the proverbially "unlucky Smith" was ordered to report at the office of the Admiral Commanding, and he had a sharp struggle to maintain a becoming composure when he heard the terse compliment and the mention of a recommendation from that austere officer, coupled with the intelligence that the zeppelin had dropped into the sea off the coast of Norway.

The spell was broken, and the brisk step and gleam in his dark eyes told their own tale as he walked quickly back to his ship.

Chapter 26

On the Sea Flank of the Allied Armies

It is a mere truism to say that the sea outflanks all land operations in warfare. Yet how many people fully realise that the left wing of the Allied armies in Belgium and France depended for its safety on the naval command of the North Sea and English Channel? Had this sea flank been permanently penetrated or forced back by the German fleet, the result must have been disastrous to a large section of the Allied military line, which actually extended from the North Sea to the Mediterranean.

Although the security of the North Sea flank did not entirely depend upon the naval forces based on Dover, Dunkirk and Harwich—as all operations, whether on land or sea, were overshadowed by the unchallenged might of the Grand Fleet, which hemmed in the entire German navy—it was upon these light forces, largely composed of units of the new navy, that the brunt of the intermittent flank fighting and the repeated attempts by the enemy to break through—with the aid of all kinds of ruses and weapons—was borne for four and a half historic years.

The detailed story of their work on the Belgian coast and in the Straits of Dover could only be told in a separate volume, but the following account of a bombardment and its sequel may not be without interest here. Its relevance to anti-submarine warfare lies in the fact that the bombardment was carried out with the object of destroying the nests of these under-water craft established in and around Zeebrugge. Much that has also been said in former chapters bases its claim to inclusion in this book almost entirely on the fact that although it did not deal exclusively with submarine fighting or minesweeping, it nevertheless formed part of the daily operations of the anti-submarine fleets, and no account of their work would bear any resemblance to the actual truth in which such seemingly extraneous episodes were excluded as irrelevant.

The Bombardment and its Sequel

There was a flat calm, with the freshness of early summer in the air. Zeebrugge lay away in the darkness some fifteen miles to the south-east—awake, watchful, but unsuspecting—when the British bombarding squadron steamed in towards the coast to take up its allotted position and wait for daybreak.

It was a heterogeneous fleet, screened by fast-moving destroyers, torpedo-boats, trawlers, M.L.'s and C.M.B.'s. The great hulls of monitors loomed black against the paling east, and the long thin lines of destroyers moved stealthily across the shadowy sea. No lights were visible, and only the occasional rhythmic thud of propellers and the call of an awakened sea-bird broke the stillness of the morning calm.

The sky was not yet alive with the whir of seaplanes, and the air remained undisturbed by the shattering roar of guns and shells. It was that brief space of time in which even Nature seems to hold her breath and make ready for the coming storm. The only movement other than the continued circling of destroyers was towards the shallow water close inshore, where powerful tugs were towing large barges—flat-bottomed craft carrying gigantic tripods made of railway metals. At predetermined places these were dropped overboard into the shallow sea and, with their legs embedded in the sandy bottom and their apices towering high above the surface, they formed observation platforms from which, in conjunction with aerial scouts, the fire of the big ships could be accurately directed on to the fortifications ashore.

These tripods were laid a distance apart and quite away from the bombarding ships, but a system of range-finding and signalling had been organised and an officer chosen as a "spotter" in each trestle.

The post of honour was on one or other of these observation towers, alone with the necessary instruments. The big shells from the shore batteries would scream overhead; some would plough up the water close by, smothering the tripod with spray, and the smaller guns would direct their fire against these eyes of the bombarding fleet. The chances were in favour of a hit, then there would be nothing left of the tripod or the spotter, simply a brief report to the Admiral Commanding that No.— observation post had been destroyed and later a fresh name in the casualty lists. It was, however, accepted as the fortune of war, and many volunteered.

The sky brightened until a pale yellow glow suffused the east, while behind the bombarding fleet the western horizon was still a cold, hazy blue. A flight of seaplanes buzzed overhead and a few minutes later the dull reports of anti-aircraft guns echoed across the miles of still water. Tiny bright flashes from white puffs of smoke appeared in the central blue, and then having got the range the great guns of the monitors roared away their charges and the scream of shells filled the air. The calm of the morning vanished, and with it the oppressive silence which precedes a battle.

It was some time before the German airmen could rise from the ground and evade the British fighting formations. In the meantime a rain of heavy projectiles from the fleet was destroying all that was destroyable of the harbour and works of Zeebrugge. With the aid of glasses huge clouds of smoke and sand could be seen rising into the air almost every second. Objects discernible one minute had disappeared when the smoke cloud of bursting shells had moved to another point of concentration a short time later. When at last the enemy's planes, in isolated ones and twos, succeeded in hovering over the fleet the surface of the sea was almost instantly broken by great spouts of white water, at first far away, then nearer, and the battle commenced in earnest.

A vast cloud of smoke now hung like a black curtain between the fleet and the shore. The M.L.'s were emitting their smoke screen to cover the bombarding ships. Shells splashed into the sea all around. The noise and vibration of the air seemed to bruise the senses, and lurid flashes came from the smoking monitors.

It was at this stage of the bombardment that the curious and unexpected happened. A white wave raced along the surface towards a monitor. It was too big for the wake of a torpedo and quite unlike the periscope of a submarine. The small, quick-firing guns of all the ships within range were trained on it and the sea around was ploughed up with shell. The white wave swerved to avoid the tornado of shot, but continued to make direct for the hull of the great floating fort at a considerable speed. Then, as it drew *very* near to its objective, a shell went home and the sea was rent by the force of a gigantic explosion, eclipsing that of any known weapon of sea warfare.

It was, however, soon discovered that the mysterious wave came from a fast torpedo-shaped boat which was evidently being controlled by electric impulses from a shore wireless station some twelve to fourteen miles distant, the necessary information regarding direction of attack being transmitted by means of wireless signals from a seaplane hovering overhead, the abnormal force of the explosion being due to the heavy charge of high explosive which such a craft was able to carry in her bow, so arranged as to fire on striking the object of attack.

With the failure of this ingenious but costly method of attack precautions were at once taken against a repetition and the seaplane hovering inconveniently overhead was driven off. The bombardment was carried on for the allotted span, by which time the shore batteries that still remained in action had found the range, notwithstanding the heavy smoke screen emitted by the M.L.'s. "Heavies" were ploughing up the water unpleasantly close to the monitors, one of which was struck, though but little damaged.

It was now considered time to draw off seawards, and the spotting officers, perched on their tripods, had to climb down the railway irons under a heavy fire and swim to the ships sent to rescue them. The tripods were then pulled over on to their sides by ropes attached to their summits and left lying in the shallow water. Under cover of the smoke screen the bombarding fleet withdrew, after inflicting severe damage on the submarine base of Zeebrugge.

****

Some two weeks previous to this bombardment a warship patrolling off the Belgian coast had reported a curious explosion in the direction of Nieuport. The night was dark and the stillness of summer rested over the Pas-de-Calais. Waves lapped gently the distant sand-dunes and war seemed a thing far away, remote as the icy winds which blow around the Poles.

In the conning-tower and at the gun stations both officers and men watched keenly, silently, for the predatory Hun. At any moment the thin blackish-brown hulls of a raiding flotilla from the bases at Zeebrugge and Ostend might slide out of the blueness of the night. The beams of searchlights would momentarily cross and recross the intervening sea and then the guns would mingle their sharp reports with the groans of dying men.

To the nerve-racking duties of night patrol in the Straits of Dover they had grown accustomed—indifferent with the contempt born of familiarity—but this did not cause any relaxation of vigilance. The element of surprise is too important a factor in modern war to be treated lightly.

So it happened that when, shortly after eight bells in the middle watch, a momentary flash of lurid flame stabbed the darkness away over the Belgian coast, and was followed by the rumble of a great but distant explosion, no one stood on his head or lost his breath blowing up a patent waistcoat, but all remained at the "still." Minutes passed and nothing happened. Slowly the destroyer crept closer inshore, but the night was dark and no further sound broke its stillness.

For two hours she scouted and listened. Little more than five miles away lay the German lines, and the theory was that somewhere in that maze of trenches and batteries an explosion had occurred.

Next day the mystery deepened, for it became known that a large portion of Nieuport Pier had been blown away during the night. As this little seaport was, however, inside the German lines, the mystery remained unexplained until after the bombardment of Zeebrugge, when it became known, in *divers* manner, that one of the electrically controlled boats had been out on a night manoeuvre and, owing to the difficulties of seaplane observation in the dark, had accidentally struck the breakwater of Nieuport.

****

Many of the patrol boats guarding the Straits of Dover or minesweeping under the fire of German coast batteries off the Belgian sand-dunes spent their days or nights of rest in the French seaport of Dunkirk, returning to Dover only after considerable periods of work on the opposite coast.

It may be thought that there was but little difference between life in the British port and that in the French town, considering the short stretch of sea between them. The following account of a night in Dunkirk will, however, give some idea of the advantage gained by having even thirty miles of blue water between an active enemy and a comfortable bed.

A NIGHT IN DUNKIRK

The night seemed uncannily quiet. In time of peace it would have passed unnoticed as just ideal summer weather, but when the human ear had grown accustomed to the almost perpetual thunder of the Flanders guns any cessation of the noise gave a feeling of disquietude, only to be likened to the hush of great forests before a tropical storm. The little town of Dunkirk, with its many ruins, was bathed in shadow, unrelieved by any artificial light, but the narrow, tortuous harbour showed a silvery streak in the brilliant moon-rays. Above the sleeping town, with its *Poilu* sentries and English sailors, was the deep indigo sky, spangled with stars.

Custom had taught the few civilian and the many naval and military inhabitants of Dunkirk to regard calm moonlight nights with very mixed feelings. It was seldom indeed that the Boche neglected such an opportunity for an air raid. Not merely one brief bombardment from the skies, but a succession of them, lasting from dusk until early morning, and repeated night after night while the weather remained favourable.

Owing to adequate preparations for such attacks the casualties were generally few, but the loss of sleep was nearly always great, unless the individual was so tired with the day's or week's minesweeping, spell in the trenches, or sea patrol that the "popping" of guns and the thud of bombs merely caused a semi-return to consciousness, with a mild, indefinable feeling of vexation at being momentarily disturbed.

To the majority, however, it meant not only the loss of sorely needed sleep, but also hard work under trying conditions. To realise fully what it is to be deprived of rest when the brain is reeling and the movement of every limb is an agony, it is necessary to have worked, marched and fought for days and nights incessantly, and then the *moral* as distinct from the *material* effect of successive air raids will be duly appreciated by those fortunate ones who spent the years 1914 to 1918 remote from the menace.

Although Dunkirk on this particular August night seemed uncannily quiet, the hour was not late. By Greenwich time it was but a few minutes past nine, and two bells had only just sounded through the many and diverse ships lying in tiers alongside the quays. So warm were the soft summer zephyrs, which scarcely stirred the surface of the water, that on the decks of many of these war-worn sweepers and patrols men lay stretched out under the sky in the sound sleep of exhaustion, while on the quays and at other points in the half-wrecked town steel-helmeted French sentries kept watch.

Of the British naval forces based on this little French seaport few were ashore, as, without special permission, both officers and men had to remain on their ships after sunset, and those not playing cards or reading in the cabins were lounging and smoking on deck. Blot out of the view the ruined houses, the shell-holes in the streets, the guns, the dug-outs and the sentries, and few scenes more unlike the popular conception of a big war base, with the enemy only a few miles distant, can be imagined.

But Dunkirk in that year of grace, 1917, did not always wear so peaceful a garb. There were frequent periods when the shells whistled over or on to the town, when the earth trembled from the concussion of high explosives, when buildings collapsed or went heavenwards in clouds of dust, when the streets were illumined with the yellow flash of *picric* acid, or were filled with clouds of poisoned gas, when ambulances clattered over the cobblestones, trains of wounded rolled in from the firing line and the killed and maimed were landed from the sea.

The first indication of the change from calm to storm came at the early hour of 10 p.m., when the air raid warning sounded throughout the town. On the quayside all was ordered haste. Mooring ropes were cast off with a minimum of shouting, and the larger ships moved slowly down the harbour towards the open sea. The few small vessels left seemed to crouch under the dock walls.

Sentries left their posts to take shelter in the great dug-outs, constructed of heavy timbers and sand-bags. These were situated at convenient points throughout the battered little town. In the houses some people descended to the cellars, but many remained wherever they happened to be, while in the cabins of the few ships which remained in harbour the games, the reading, the letter-writing and, in a few cases, even the sleeping went on undisturbed.

After a short interval of oppressive silence, during which time no light or sound came from the seemingly deserted town, a faint whir of propellers became just audible in the stillness of the summer night. Then it died away momentarily. Suddenly a bright glare, like that of a star-shell, lit up the roofs and streets, and almost simultaneously came the dull vibrating report of a bomb. It sounded from the direction of the cathedral. Searchlights flashed out from various points, but their powerful rays were lost in the luminous vault above. Guns roared and bright flashes appeared like summer lightning in the sky. Every few seconds the town trembled from the shock of exploding bombs, first at one point and then at another, but nothing could be seen of the raiding squadron. Pieces from the shells bursting overhead and fragments of bombs and shattered masonry fell like rain into the streets and into the waters of the harbour.

On the quayside a big aerial torpedo had made a crater large enough to bury the horse which it had killed in a near-by stable.

A few seconds later another bomb fell close to a minesweeper and a fragment gashed the decks but did not penetrate them. In the cabins the concussion of almost every bomb which fell on shore was felt with curious precision. The glass of wheel-houses and deck cabins was shattered, and the rattle and thud on the decks and iron sides denoted the storm of falling metal.

The din of the raid went on for some time and then died away with a final long-range shot from "Loose Lizzie" on the hills behind. When all was clear heads appeared from hatchways, dug-outs and cellars. People searched the sky curiously in an endeavour to make sure that there was "no deception," although from first to last nothing had been seen of the raiders except by those with the instruments, the searchlights and the guns. The latest news of the damage caused—two houses, a man and a horse—went from mouth to mouth. Then the summer night regained its tranquillity and Dunkirk slept.

****

The familiar boom sounded its loudest in the stillness of the night and the ground seemed to tremble the more violently because of the darkness. It was 1 a.m. The young moon had sunk beneath the horizon and a light film of cloud had drifted over the sky.

The old French reservist doing sentry-go on the quay glanced up with a shrug of indifference and slowly shouldering his rifle walked leisurely towards a dug-out. Searchlights became busy exploring the sky. This time their rays were not lost in the opaque blueness above, but went up in well-defined columns of light until reflected on the lofty clouds. Presently the beams concentrated and, when the eyes had grown accustomed to the glare, little white "butterflies" were seen circling in the upper air. Then the guns opened fire and white puffs, like tiny balls of cotton-wool, appeared among the butterflies. The earth trembled with the explosion of falling bombs and the recoil of anti-aircraft batteries. A little flicker of yellow light appeared in the circle of white. The guns increased in violence. The yellow light grew in size. It was falling. The burning machine crashed to earth.

The bombs and the gun-fire lasted for some twenty minutes and then ceased suddenly, as if by prearranged signal. Allied squadrons were in the air and the distant crackle of machine guns sounded from the skies. It died away, however, almost immediately, but the raiders were chased back to within their own lines minus two of their number.

With the coming of dawn two solitary hostile machines circling at a fairly low altitude could be seen. They dropped no bombs, but the reason for their presence was soon apparent. Shells from the long-range guns behind the German lines began to moan, whistle and burst in and around the luckless town. A hit was signified by a cloud of smoke, dust and debris, and ambulances again became busy in the stone-paved streets.

One shell, carrying sufficient explosive to blow up an average-sized ship, ploughed up the water of the harbour, but did no damage, and by 6 a.m. Allied squadrons had chased away the hostile aerial observers. Once again the peace of an ideal summer morning reigned over the historic town.

The few minesweeping and other ships which had remained in the harbour through the night now commenced to show signs of returning life and activity. Heavy brown smoke poured from the funnels of some, the staccato noise of oil engines came from others, and men were busy on the decks of all. The night's "rest" was over and the vital work of sweeping, possibly under an irritating fire from shore batteries and the strain of a necessarily ever-alert patrol, commenced afresh. The steady barometer promised a fine day for the harvesting of mines and, for the ships that returned, another night's *rest* similar to the previous three!

LEONAUR

ALSO FROM LEONAUR
AVAILABLE IN SOFTCOVER OR HARDCOVER WITH DUST JACKET

JOURNALS OF ROBERT ROGERS OF THE RANGERS *by Robert Rogers*—The exploits of Rogers & the Rangers in his own words during 1755-1761 in the French & Indian War.

GALLOPING GUNS *by James Young*—The Experiences of an Officer of the Bengal Horse Artillery During the Second Maratha War 1804-1805.

GORDON *by Demetrius Charles Boulger*—The Career of Gordon of Khartoum.

THE BATTLE OF NEW ORLEANS *by Zachary F. Smith*—The final major engagement of the War of 1812.

THE TWO WARS OF MRS DUBERLY *by Frances Isabella Duberly*—An Intrepid Victorian Lady's Experience of the Crimea and Indian Mutiny.

WITH THE GUARDS' BRIGADE DURING THE BOER WAR *by Edward P. Lowry*—On Campaign from Bloemfontein to Koomati Poort and Back.

THE REBELLIOUS DUCHESS *by Paul F. S. Dermoncourt*—The Adventures of the Duchess of Berri and Her Attempt to Overthrow French Monarchy.

MEN OF THE MUTINY *by John Tulloch Nash & Henry Metcalfe*—Two Accounts of the Great Indian Mutiny of 1857: Fighting with the Bengal Yeomanry Cavalry & Private Metcalfe at Lucknow.

CAMPAIGN IN THE CRIMEA *by George Shuldham Peard*—The Recollections of an Officer of the 20th Regiment of Foot.

WITHIN SEBASTOPOL *by K. Hodasevich*—A Narrative of the Campaign in the Crimea, and of the Events of the Siege.

WITH THE CAVALRY TO AFGHANISTAN *by William Taylor*—The Experiences of a Trooper of H. M. 4th Light Dragoons During the First Afghan War.

THE CAWNPORE MAN *by Mowbray Thompson*—A First Hand Account of the Siege and Massacre During the Indian Mutiny By One of Four Survivors.

BRIGADE COMMANDER: AFGHANISTAN *by Henry Brooke*—The Journal of the Commander of the 2nd Infantry Brigade, Kandahar Field Force During the Second Afghan War.

BANCROFT OF THE BENGAL HORSE ARTILLERY *by N. W. Bancroft*—An Account of the First Sikh War 1845-1846.

AVAILABLE ONLINE AT www.leonaur.com
AND FROM ALL GOOD BOOK STORES

07/09

LEONAUR

ALSO FROM LEONAUR
AVAILABLE IN SOFTCOVER OR HARDCOVER WITH DUST JACKET

AFGHANISTAN: THE BELEAGUERED BRIGADE *by G. R. Gleig*—An Account of Sale's Brigade During the First Afghan War.

IN THE RANKS OF THE C. I. V *by Erskine Childers*—With the City Imperial Volunteer Battery (Honourable Artillery Company) in the Second Boer War.

THE BENGAL NATIVE ARMY *by F. G. Cardew*—An Invaluable Reference Resource.

THE 7TH (QUEEN'S OWN) HUSSARS *by C. R. B. Barrett*—Uniforms, Equipment, Weapons, Traditions, the Services of Notable Officers and Men & the Appendices to All Volumes—Volume 4: 1688-1914.

THE SWORD OF THE CROWN *by Eric W. Sheppard*—A History of the British Army to 1914.

THE 7TH (QUEEN'S OWN) HUSSARS *by C. R. B. Barrett*—On Campaign During the Canadian Rebellion, the Indian Mutiny, the Sudan, Matabeleland, Mashonaland and the Boer War Volume 3: 1818-1914.

THE KHARTOUM CAMPAIGN *by Bennet Burleigh*—A Special Correspondent's View of the Reconquest of the Sudan by British and Egyptian Forces under Kitchener—1898.

EL PUCHERO *by Richard McSherry*—The Letters of a Surgeon of Volunteers During Scott's Campaign of the American-Mexican War 1847-1848.

RIFLEMAN SAHIB *by E. Maude*—The Recollections of an Officer of the Bombay Rifles During the Southern Mahratta Campaign, Second Sikh War, Persian Campaign and Indian Mutiny.

THE KING'S HUSSAR *by Edwin Mole*—The Recollections of a 14th (King's) Hussar During the Victorian Era.

JOHN COMPANY'S CAVALRYMAN *by William Johnson*—The Experiences of a British Soldier in the Crimea, the Persian Campaign and the Indian Mutiny.

COLENSO & DURNFORD'S ZULU WAR *by Frances E. Colenso & Edward Durnford*—The first and possibly the most important history of the Zulu War.

U. S. DRAGOON *by Samuel E. Chamberlain*—Experiences in the Mexican War 1846-48 and on the South Western Frontier.

AVAILABLE ONLINE AT www.leonaur.com
AND FROM ALL GOOD BOOK STORES

07/09

LEONAUR

ALSO FROM LEONAUR
AVAILABLE IN SOFTCOVER OR HARDCOVER WITH DUST JACKET

THE 2ND MAORI WAR: 1860-1861 *by Robert Carey*—The Second Maori War, or First Taranaki War, one more bloody instalment of the conflicts between European settlers and the indigenous Maori people.

A JOURNAL OF THE SECOND SIKH WAR *by Daniel A. Sandford*—The Experiences of an Ensign of the 2nd Bengal European Regiment During the Campaign in the Punjab, India, 1848-49.

THE LIGHT INFANTRY OFFICER *by John H. Cooke*—The Experiences of an Officer of the 43rd Light Infantry in America During the War of 1812.

BUSHVELDT CARBINEERS *by George Witton*—The War Against the Boers in South Africa and the 'Breaker' Morant Incident.

LAKE'S CAMPAIGNS IN INDIA *by Hugh Pearse*—The Second Anglo Maratha War, 1803-1807.

BRITAIN IN AFGHANISTAN 1: THE FIRST AFGHAN WAR 1839-42 *by Archibald Forbes*—From invasion to destruction-a British military disaster.

BRITAIN IN AFGHANISTAN 2: THE SECOND AFGHAN WAR 1878-80 *by Archibald Forbes*—This is the history of the Second Afghan War-another episode of British military history typified by savagery, massacre, siege and battles.

UP AMONG THE PANDIES *by Vivian Dering Majendie*—Experiences of a British Officer on Campaign During the Indian Mutiny, 1857-1858.

MUTINY: 1857 *by James Humphries*—Authentic Voices from the Indian Mutiny-First Hand Accounts of Battles, Sieges and Personal Hardships.

BLOW THE BUGLE, DRAW THE SWORD *by W. H. G. Kingston*—The Wars, Campaigns, Regiments and Soldiers of the British & Indian Armies During the Victorian Era, 1839-1898.

WAR BEYOND THE DRAGON PAGODA *by Major J. J. Snodgrass*—A Personal Narrative of the First Anglo-Burmese War 1824 - 1826.

THE HERO OF ALIWAL *by James Humphries*—The Campaigns of Sir Harry Smith in India, 1843-1846, During the Gwalior War & the First Sikh War.

ALL FOR A SHILLING A DAY *by Donald F. Featherstone*—The story of H.M. 16th, the Queen's Lancers During the first Sikh War 1845-1846.

AVAILABLE ONLINE AT www.leonaur.com
AND FROM ALL GOOD BOOK STORES

07/09

LEONAUR

ALSO FROM LEONAUR
AVAILABLE IN SOFTCOVER OR HARDCOVER WITH DUST JACKET

THE FALL OF THE MOGHUL EMPIRE OF HINDUSTAN *by H. G. Keene*—By the beginning of the nineteenth century, as British and Indian armies under Lake and Wellesley dominated the scene, a little over half a century of conflict brought the Moghul Empire to its knees.

LADY SALE'S AFGHANISTAN *by Florentia Sale*—An Indomitable Victorian Lady's Account of the Retreat from Kabul During the First Afghan War.

THE CAMPAIGN OF MAGENTA AND SOLFERINO 1859 *by Harold Carmichael Wylly*—The Decisive Conflict for the Unification of Italy.

FRENCH'S CAVALRY CAMPAIGN *by J. G. Maydon*—A Special Corresponent's View of British Army Mounted Troops During the Boer War.

CAVALRY AT WATERLOO *by Sir Evelyn Wood*—British Mounted Troops During the Campaign of 1815.

THE SUBALTERN *by George Robert Gleig*—The Experiences of an Officer of the 85th Light Infantry During the Peninsular War.

NAPOLEON AT BAY, 1814 *by F. Loraine Petre*—The Campaigns to the Fall of the First Empire.

NAPOLEON AND THE CAMPAIGN OF 1806 *by Colonel Vachée*—The Napoleonic Method of Organisation and Command to the Battles of Jena & Auerstädt.

THE COMPLETE ADVENTURES IN THE CONNAUGHT RANGERS *by William Grattan*—The 88th Regiment during the Napoleonic Wars by a Serving Officer.

BUGLER AND OFFICER OF THE RIFLES *by William Green & Harry Smith*—With the 95th (Rifles) during the Peninsular & Waterloo Campaigns of the Napoleonic Wars.

NAPOLEONIC WAR STORIES *by Sir Arthur Quiller-Couch*—Tales of soldiers, spies, battles & sieges from the Peninsular & Waterloo campaingns.

CAPTAIN OF THE 95TH (RIFLES) *by Jonathan Leach*—An officer of Wellington's sharpshooters during the Peninsular, South of France and Waterloo campaigns of the Napoleonic wars.

RIFLEMAN COSTELLO *by Edward Costello*—The adventures of a soldier of the 95th (Rifles) in the Peninsular & Waterloo Campaigns of the Napoleonic wars.

AVAILABLE ONLINE AT **www.leonaur.com**
AND FROM ALL GOOD BOOK STORES
07/09

ALSO FROM LEONAUR
AVAILABLE IN SOFTCOVER OR HARDCOVER WITH DUST JACKET

AT THEM WITH THE BAYONET *by Donald F. Featherstone*—The first Anglo-Sikh War 1845-1846.

STEPHEN CRANE'S BATTLES *by Stephen Crane*—Nine Decisive Battles Recounted by the Author of 'The Red Badge of Courage'.

THE GURKHA WAR *by H. T. Prinsep*—The Anglo-Nepalese Conflict in North East India 1814-1816.

FIRE & BLOOD *by G. R. Gleig*—The burning of Washington & the battle of New Orleans, 1814, through the eyes of a young British soldier.

SOUND ADVANCE! *by Joseph Anderson*—Experiences of an officer of HM 50th regiment in Australia, Burma & the Gwalior war.

THE CAMPAIGN OF THE INDUS *by Thomas Holdsworth*—Experiences of a British Officer of the 2nd (Queen's Royal) Regiment in the Campaign to Place Shah Shuja on the Throne of Afghanistan 1838 - 1840.

WITH THE MADRAS EUROPEAN REGIMENT IN BURMA *by John Butler*—The Experiences of an Officer of the Honourable East India Company's Army During the First Anglo-Burmese War 1824 - 1826.

IN ZULULAND WITH THE BRITISH ARMY *by Charles L. Norris-Newman*—The Anglo-Zulu war of 1879 through the first-hand experiences of a special correspondent.

BESIEGED IN LUCKNOW *by Martin Richard Gubbins*—The first Anglo-Sikh War 1845-1846.

A TIGER ON HORSEBACK *by L. March Phillips*—The Experiences of a Trooper & Officer of Rimington's Guides - The Tigers - during the Anglo-Boer war 1899 - 1902.

SEPOYS, SIEGE & STORM *by Charles John Griffiths*—The Experiences of a young officer of H.M.'s 61st Regiment at Ferozepore, Delhi ridge and at the fall of Delhi during the Indian mutiny 1857.

CAMPAIGNING IN ZULULAND *by W. E. Montague*—Experiences on campaign during the Zulu war of 1879 with the 94th Regiment.

THE STORY OF THE GUIDES *by G.J. Younghusband*—The Exploits of the Soldiers of the famous Indian Army Regiment from the northwest frontier 1847 - 1900.

AVAILABLE ONLINE AT www.leonaur.com
AND FROM ALL GOOD BOOK STORES

07/09

LEONAUR

ALSO FROM LEONAUR
AVAILABLE IN SOFTCOVER OR HARDCOVER WITH DUST JACKET

ZULU:1879 *by D.C.F. Moodie & the Leonaur Editors*—The Anglo-Zulu War of 1879 from contemporary sources: First Hand Accounts, Interviews, Dispatches, Official Documents & Newspaper Reports.

THE RED DRAGOON *by W.J. Adams*—With the 7th Dragoon Guards in the Cape of Good Hope against the Boers & the Kaffir tribes during the 'war of the axe' 1843-48'.

THE RECOLLECTIONS OF SKINNER OF SKINNER'S HORSE *by James Skinner*—James Skinner and his 'Yellow Boys' Irregular cavalry in the wars of India between the British, Mahratta, Rajput, Mogul, Sikh & Pindarree Forces.

A CAVALRY OFFICER DURING THE SEPOY REVOLT *by A. R. D. Mackenzie*—Experiences with the 3rd Bengal Light Cavalry, the Guides and Sikh Irregular Cavalry from the outbreak to Delhi and Lucknow.

A NORFOLK SOLDIER IN THE FIRST SIKH WAR *by J W Baldwin*—Experiences of a private of H.M. 9th Regiment of Foot in the battles for the Punjab, India 1845-6.

TOMMY ATKINS' WAR STORIES: 14 FIRST HAND ACCOUNTS—Fourteen first hand accounts from the ranks of the British Army during Queen Victoria's Empire.

THE WATERLOO LETTERS *by H. T. Siborne*—Accounts of the Battle by British Officers for its Foremost Historian.

NEY: GENERAL OF CAVALRY VOLUME 1—1769-1799 *by Antoine Bulos*—The Early Career of a Marshal of the First Empire.

NEY: MARSHAL OF FRANCE VOLUME 2—1799-1805 *by Antoine Bulos*—The Early Career of a Marshal of the First Empire.

AIDE-DE-CAMP TO NAPOLEON *by Philippe-Paul de Ségur*—For anyone interested in the Napoleonic Wars this book, written by one who was intimate with the strategies and machinations of the Emperor, will be essential reading.

TWILIGHT OF EMPIRE *by Sir Thomas Ussher & Sir George Cockburn*—Two accounts of Napoleon's Journeys in Exile to Elba and St. Helena: Narrative of Events by Sir Thomas Ussher & Napoleon's Last Voyage: Extract of a diary by Sir George Cockburn.

PRIVATE WHEELER *by William Wheeler*—The letters of a soldier of the 51st Light Infantry during the Peninsular War & at Waterloo.

AVAILABLE ONLINE AT **www.leonaur.com**
AND FROM ALL GOOD BOOK STORES

07/09

LEONAUR

ALSO FROM LEONAUR
AVAILABLE IN SOFTCOVER OR HARDCOVER WITH DUST JACKET

OFFICERS & GENTLEMEN *by Peter Hawker & William Graham*—Two Accounts of British Officers During the Peninsula War: Officer of Light Dragoons by Peter Hawker & Campaign in Portugal and Spain by William Graham .

THE WALCHEREN EXPEDITION *by Anonymous*—The Experiences of a British Officer of the 81st Regt. During the Campaign in the Low Countries of 1809.

LADIES OF WATERLOO *by Charlotte A. Eaton, Magdalene de Lancey & Juana Smith*—The Experiences of Three Women During the Campaign of 1815: Waterloo Days by Charlotte A. Eaton, A Week at Waterloo by Magdalene de Lancey & Juana's Story by Juana Smith.

JOURNAL OF AN OFFICER IN THE KING'S GERMAN LEGION *by John Frederick Hering*—Recollections of Campaigning During the Napoleonic Wars.

JOURNAL OF AN ARMY SURGEON IN THE PENINSULAR WAR *by Charles Boutflower*—The Recollections of a British Army Medical Man on Campaign During the Napoleonic Wars.

ON CAMPAIGN WITH MOORE AND WELLINGTON *by Anthony Hamilton*—The Experiences of a Soldier of the 43rd Regiment During the Peninsular War.

THE ROAD TO AUSTERLITZ *by R. G. Burton*—Napoleon's Campaign of 1805.

SOLDIERS OF NAPOLEON *by A. J. Doisy De Villargennes & Arthur Chuquet*—The Experiences of the Men of the French First Empire: Under the Eagles by A. J. Doisy De Villargennes & Voices of 1812 by Arthur Chuquet .

INVASION OF FRANCE, 1814 *by F. W. O. Maycock*—The Final Battles of the Napoleonic First Empire.

LEIPZIG—A CONFLICT OF TITANS *by Frederic Shoberl*—A Personal Experience of the 'Battle of the Nations' During the Napoleonic Wars, October 14th-19th, 1813.

SLASHERS *by Charles Cadell*—The Campaigns of the 28th Regiment of Foot During the Napoleonic Wars by a Serving Officer.

BATTLE IMPERIAL *by Charles William Vane*—The Campaigns in Germany & France for the Defeat of Napoleon 1813-1814.

SWIFT & BOLD *by Gibbes Rigaud*—The 60th Rifles During the Peninsula War.

AVAILABLE ONLINE AT www.leonaur.com
AND FROM ALL GOOD BOOK STORES

07/09

LEONAUR

ALSO FROM LEONAUR
AVAILABLE IN SOFTCOVER OR HARDCOVER WITH DUST JACKET

ADVENTURES OF A YOUNG RIFLEMAN *by Johann Christian Maempel*—The Experiences of a Saxon in the French & British Armies During the Napoleonic Wars.

THE HUSSAR *by Norbert Landsheit & G. R. Gleig*—A German Cavalryman in British Service Throughout the Napoleonic Wars.

RECOLLECTIONS OF THE PENINSULA *by Moyle Sherer*—An Officer of the 34th Regiment of Foot—'The Cumberland Gentlemen'—on Campaign Against Napoleon's French Army in Spain.

MARINE OF REVOLUTION & CONSULATE *by Moreau de Jonnès*—The Recollections of a French Soldier of the Revolutionary Wars 1791-1804.

GENTLEMEN IN RED *by John Dobbs & Robert Knowles*—Two Accounts of British Infantry Officers During the Peninsular War Recollections of an Old 52nd Man by John Dobbs An Officer of Fusiliers by Robert Knowles.

CORPORAL BROWN'S CAMPAIGNS IN THE LOW COUNTRIES *by Robert Brown*—Recollections of a Coldstream Guard in the Early Campaigns Against Revolutionary France 1793-1795.

THE 7TH (QUEENS OWN) HUSSARS *by C. R. B. Barrett*—During the Campaigns in the Low Countries & the Peninsula and Waterloo Campaigns of the Napoleonic Wars. Volume 2: 1793-1815.

THE MARENGO CAMPAIGN 1800 *by Herbert H. Sargent*—The Victory that Completed the Austrian Defeat in Italy.

DONALDSON OF THE 94TH—SCOTS BRIGADE *by Joseph Donaldson*—The Recollections of a Soldier During the Peninsula & South of France Campaigns of the Napoleonic Wars.

A CONSCRIPT FOR EMPIRE *by Philippe as told to Johann Christian Maempel*—The Experiences of a Young German Conscript During the Napoleonic Wars.

JOURNAL OF THE CAMPAIGN OF 1815 *by Alexander Cavalié Mercer*—The Experiences of an Officer of the Royal Horse Artillery During the Waterloo Campaign.

NAPOLEON'S CAMPAIGNS IN POLAND 1806-7 *by Robert Wilson*—The campaign in Poland from the Russian side of the conflict.

AVAILABLE ONLINE AT www.leonaur.com
AND FROM ALL GOOD BOOK STORES

07/09

LEONAUR

ALSO FROM LEONAUR
AVAILABLE IN SOFTCOVER OR HARDCOVER WITH DUST JACKET

OMPTEDA OF THE KING'S GERMAN LEGION *by Christian von Ompteda*—A Hanoverian Officer on Campaign Against Napoleon.

LIEUTENANT SIMMONS OF THE 95TH (RIFLES) *by George Simmons*—Recollections of the Peninsula, South of France & Waterloo Campaigns of the Napoleonic Wars.

A HORSEMAN FOR THE EMPEROR *by Jean Baptiste Gazzola*—A Cavalryman of Napoleon's Army on Campaign Throughout the Napoleonic Wars.

SERGEANT LAWRENCE *by William Lawrence*—With the 40th Regt. of Foot in South America, the Peninsular War & at Waterloo.

CAMPAIGNS WITH THE FIELD TRAIN *by Richard D. Henegan*—Experiences of a British Officer During the Peninsula and Waterloo Campaigns of the Napoleonic Wars.

CAVALRY SURGEON *by S. D. Broughton*—On Campaign Against Napoleon in the Peninsula & South of France During the Napoleonic Wars 1812-1814.

MEN OF THE RIFLES *by Thomas Knight, Henry Curling & Jonathan Leach*—The Reminiscences of Thomas Knight of the 95th (Rifles) by Thomas Knight, Henry Curling's Anecdotes by Henry Curling & The Field Services of the Rifle Brigade from its Formation to Waterloo by Jonathan Leach.

THE ULM CAMPAIGN 1805 *by F. N. Maude*—Napoleon and the Defeat of the Austrian Army During the 'War of the Third Coalition'.

SOLDIERING WITH THE 'DIVISION' *by Thomas Garrety*—The Military Experiences of an Infantryman of the 43rd Regiment During the Napoleonic Wars.

SERGEANT MORRIS OF THE 73RD FOOT *by Thomas Morris*—The Experiences of a British Infantryman During the Napoleonic Wars-Including Campaigns in Germany and at Waterloo.

A VOICE FROM WATERLOO *by Edward Cotton*—The Personal Experiences of a British Cavalryman Who Became a Battlefield Guide and Authority on the Campaign of 1815.

NAPOLEON AND HIS MARSHALS *by J. T. Headley*—The Men of the First Empire.

AVAILABLE ONLINE AT **www.leonaur.com**
AND FROM ALL GOOD BOOK STORES

07/09

LEONAUR

ALSO FROM LEONAUR
AVAILABLE IN SOFTCOVER OR HARDCOVER WITH DUST JACKET

COLBORNE: A SINGULAR TALENT FOR WAR *by John Colborne*—The Napoleonic Wars Career of One of Wellington's Most Highly Valued Officers in Egypt, Holland, Italy, the Peninsula and at Waterloo.

NAPOLEON'S RUSSIAN CAMPAIGN *by Philippe Henri de Segur*—The Invasion, Battles and Retreat by an Aide-de-Camp on the Emperor's Staff.

WITH THE LIGHT DIVISION *by John H. Cooke*—The Experiences of an Officer of the 43rd Light Infantry in the Peninsula and South of France During the Napoleonic Wars.

WELLINGTON AND THE PYRENEES CAMPAIGN VOLUME I: FROM VITORIA TO THE BIDASSOA *by F. C. Beatson*—The final phase of the campaign in the Iberian Peninsula.

WELLINGTON AND THE INVASION OF FRANCE VOLUME II: THE BIDASSOA TO THE BATTLE OF THE NIVELLE *by F. C. Beatson*—The final phase of the campaign in the Iberian Peninsula.

WELLINGTON AND THE FALL OF FRANCE VOLUME III: THE GAVES AND THE BATTLE OF ORTHEZ *by F. C. Beatson*—The final phase of the campaign in the Iberian Peninsula.

NAPOLEON'S IMPERIAL GUARD: FROM MARENGO TO WATERLOO *by J. T. Headley*—The story of Napoleon's Imperial Guard and the men who commanded them.

BATTLES & SIEGES OF THE PENINSULAR WAR *by W. H. Fitchett*—Corunna, Busaco, Albuera, Ciudad Rodrigo, Badajos, Salamanca, San Sebastian & Others.

SERGEANT GUILLEMARD: THE MAN WHO SHOT NELSON? *by Robert Guillemard*—A Soldier of the Infantry of the French Army of Napoleon on Campaign Throughout Europe.

WITH THE GUARDS ACROSS THE PYRENEES *by Robert Batty*—The Experiences of a British Officer of Wellington's Army During the Battles for the Fall of Napoleonic France, 1813 .

A STAFF OFFICER IN THE PENINSULA *by E. W. Buckham*—An Officer of the British Staff Corps Cavalry During the Peninsula Campaign of the Napoleonic Wars.

THE LEIPZIG CAMPAIGN: 1813—NAPOLEON AND THE "BATTLE OF THE NATIONS" *by F. N. Maude*—Colonel Maude's analysis of Napoleon's campaign of 1813 around Leipzig.

AVAILABLE ONLINE AT **www.leonaur.com**
AND FROM ALL GOOD BOOK STORES
07/09

LEONAUR

ALSO FROM LEONAUR
AVAILABLE IN SOFTCOVER OR HARDCOVER WITH DUST JACKET

BUGEAUD: A PACK WITH A BATON *by Thomas Robert Bugeaud*—The Early Campaigns of a Soldier of Napoleon's Army Who Would Become a Marshal of France.

WATERLOO RECOLLECTIONS *by Frederick Llewellyn*—Rare First Hand Accounts, Letters, Reports and Retellings from the Campaign of 1815.

SERGEANT NICOL *by Daniel Nicol*—The Experiences of a Gordon Highlander During the Napoleonic Wars in Egypt, the Peninsula and France.

THE JENA CAMPAIGN: 1806 *by F. N. Maude*—The Twin Battles of Jena & Auerstadt Between Napoleon's French and the Prussian Army.

PRIVATE O'NEIL *by Charles O'Neil*—The recollections of an Irish Rogue of H. M. 28th Regt.—The Slashers—during the Peninsula & Waterloo campaigns of the Napoleonic war.

ROYAL HIGHLANDER *by James Anton*—A soldier of H.M 42nd (Royal) Highlanders during the Peninsular, South of France & Waterloo Campaigns of the Napoleonic Wars.

CAPTAIN BLAZE *by Elzéar Blaze*—Life in Napoleons Army.

LEJEUNE VOLUME 1 *by Louis-François Lejeune*—The Napoleonic Wars through the Experiences of an Officer on Berthier's Staff.

LEJEUNE VOLUME 2 *by Louis-François Lejeune*—The Napoleonic Wars through the Experiences of an Officer on Berthier's Staff.

CAPTAIN COIGNET *by Jean-Roch Coignet*—A Soldier of Napoleon's Imperial Guard from the Italian Campaign to Russia and Waterloo.

FUSILIER COOPER *by John S. Cooper*—Experiences in the 7th (Royal) Fusiliers During the Peninsular Campaign of the Napoleonic Wars and the American Campaign to New Orleans.

FIGHTING NAPOLEON'S EMPIRE *by Joseph Anderson*—The Campaigns of a British Infantryman in Italy, Egypt, the Peninsular & the West Indies During the Napoleonic Wars.

CHASSEUR BARRES *by Jean-Baptiste Barres*—The experiences of a French Infantryman of the Imperial Guard at Austerlitz, Jena, Eylau, Friedland, in the Peninsular, Lutzen, Bautzen, Zinnwald and Hanau during the Napoleonic Wars.

AVAILABLE ONLINE AT **www.leonaur.com**
AND FROM ALL GOOD BOOK STORES
07/09

LEONAUR

ALSO FROM LEONAUR
AVAILABLE IN SOFTCOVER OR HARDCOVER WITH DUST JACKET

CAPTAIN COIGNET *by Jean-Roch Coignet*—A Soldier of Napoleon's Imperial Guard from the Italian Campaign to Russia and Waterloo.

HUSSAR ROCCA *by Albert Jean Michel de Rocca*—A French cavalry officer's experiences of the Napoleonic Wars and his views on the Peninsular Campaigns against the Spanish, British And Guerilla Armies.

MARINES TO 95TH (RIFLES) *by Thomas Fernyhough*—The military experiences of Robert Fernyhough during the Napoleonic Wars.

LIGHT BOB *by Robert Blakeney*—The experiences of a young officer in H.M 28th & 36th regiments of the British Infantry during the Peninsular Campaign of the Napoleonic Wars 1804 - 1814.

WITH WELLINGTON'S LIGHT CAVALRY *by William Tomkinson*—The Experiences of an officer of the 16th Light Dragoons in the Peninsular and Waterloo campaigns of the Napoleonic Wars.

SERGEANT BOURGOGNE *by Adrien Bourgogne*—With Napoleon's Imperial Guard in the Russian Campaign and on the Retreat from Moscow 1812 - 13.

SURTEES OF THE 95TH (RIFLES) *by William Surtees*—A Soldier of the 95th (Rifles) in the Peninsular campaign of the Napoleonic Wars.

SWORDS OF HONOUR *by Henry Newbolt & Stanley L. Wood*—The Careers of Six Outstanding Officers from the Napoleonic Wars, the Wars for India and the American Civil War.

ENSIGN BELL IN THE PENINSULAR WAR *by George Bell*—The Experiences of a young British Soldier of the 34th Regiment 'The Cumberland Gentlemen' in the Napoleonic wars.

HUSSAR IN WINTER *by Alexander Gordon*—A British Cavalry Officer during the retreat to Corunna in the Peninsular campaign of the Napoleonic Wars.

THE COMPLEAT RIFLEMAN HARRIS *by Benjamin Harris as told to and transcribed by Captain Henry Curling, 52nd Regt. of Foot*—The adventures of a soldier of the 95th (Rifles) during the Peninsular Campaign of the Napoleonic Wars.

THE ADVENTURES OF A LIGHT DRAGOON *by George Farmer & G.R. Gleig*—A cavalryman during the Peninsular & Waterloo Campaigns, in captivity & at the siege of Bhurtpore, India.

AVAILABLE ONLINE AT **www.leonaur.com**
AND FROM ALL GOOD BOOK STORES
07/09

LEONAUR

ALSO FROM LEONAUR
AVAILABLE IN SOFTCOVER OR HARDCOVER WITH DUST JACKET

THE LIFE OF THE REAL BRIGADIER GERARD VOLUME 1—THE YOUNG HUSSAR 1782-1807 *by Jean-Baptiste De Marbot*—A French Cavalryman Of the Napoleonic Wars at Marengo, Austerlitz, Jena, Eylau & Friedland.

THE LIFE OF THE REAL BRIGADIER GERARD VOLUME 2—IMPERIAL AIDE-DE-CAMP 1807-1811 *by Jean-Baptiste De Marbot*—A French Cavalryman of the Napoleonic Wars at Saragossa, Landshut, Eckmuhl, Ratisbon, Aspern-Essling, Wagram, Busaco & Torres Vedras.

THE LIFE OF THE REAL BRIGADIER GERARD VOLUME 3—COLONEL OF CHASSEURS 1811-1815 *by Jean-Baptiste De Marbot*—A French Cavalryman in the retreat from Moscow, Lutzen, Bautzen, Katzbach, Leipzig, Hanau & Waterloo.

THE INDIAN WAR OF 1864 *by Eugene Ware*—The Experiences of a Young Officer of the 7th Iowa Cavalry on the Western Frontier During the Civil War.

THE MARCH OF DESTINY *by Charles E. Young & V. Devinny*—Dangers of the Trail in 1865 by Charles E. Young & The Story of a Pioneer by V. Devinny, two Accounts of Early Emigrants to Colorado.

CROSSING THE PLAINS *by William Audley Maxwell*—A First Hand Narrative of the Early Pioneer Trail to California in 1857.

CHIEF OF SCOUTS *by William F. Drannan*—A Pilot to Emigrant and Government Trains, Across the Plains of the Western Frontier.

THIRTY-ONE YEARS ON THE PLAINS AND IN THE MOUNTAINS *by William F. Drannan*—William Drannan was born to be a pioneer, hunter, trapper and wagon train guide during the momentous days of the Great American West.

THE INDIAN WARS VOLUNTEER *by William Thompson*—Recollections of the Conflict Against the Snakes, Shoshone, Bannocks, Modocs and Other Native Tribes of the American North West.

THE 4TH TENNESSEE CAVALRY *by George B. Guild*—The Services of Smith's Regiment of Confederate Cavalry by One of its Officers.

COLONEL WORTHINGTON'S SHILOH *by T. Worthington*—The Tennessee Campaign, 1862, by an Officer of the Ohio Volunteers.

FOUR YEARS IN THE SADDLE *by W. L. Curry*—The History of the First Regiment Ohio Volunteer Cavalry in the American Civil War.

AVAILABLE ONLINE AT www.leonaur.com
AND FROM ALL GOOD BOOK STORES
07/09

![LEONAUR logo]

ALSO FROM LEONAUR
AVAILABLE IN SOFTCOVER OR HARDCOVER WITH DUST JACKET

LIFE IN THE ARMY OF NORTHERN VIRGINIA *by Carlton McCarthy*—The Observations of a Confederate Artilleryman of Cutshaw's Battalion During the American Civil War 1861-1865.

HISTORY OF THE CAVALRY OF THE ARMY OF THE POTOMAC *by Charles D. Rhodes*—Including Pope's Army of Virginia and the Cavalry Operations in West Virginia During the American Civil War.

CAMP-FIRE AND COTTON-FIELD *by Thomas W. Knox*—A New York Herald Correspondent's View of the American Civil War.

SERGEANT STILLWELL *by Leander Stillwell* —The Experiences of a Union Army Soldier of the 61st Illinois Infantry During the American Civil War.

STONEWALL'S CANNONEER *by Edward A. Moore*—Experiences with the Rockbridge Artillery, Confederate Army of Northern Virginia, During the American Civil War.

THE SIXTH CORPS *by George Stevens*—The Army of the Potomac, Union Army, During the American Civil War.

THE RAILROAD RAIDERS *by William Pittenger*—An Ohio Volunteers Recollections of the Andrews Raid to Disrupt the Confederate Railroad in Georgia During the American Civil War.

CITIZEN SOLDIER *by John Beatty*—An Account of the American Civil War by a Union Infantry Officer of Ohio Volunteers Who Became a Brigadier General.

COX: PERSONAL RECOLLECTIONS OF THE CIVIL WAR--VOLUME 1 *by Jacob Dolson Cox*—West Virginia, Kanawha Valley, Gauley Bridge, Cotton Mountain, South Mountain, Antietam, the Morgan Raid & the East Tennessee Campaign.

COX: PERSONAL RECOLLECTIONS OF THE CIVIL WAR--VOLUME 2 *by Jacob Dolson Cox*—Siege of Knoxville, East Tennessee, Atlanta Campaign, the Nashville Campaign & the North Carolina Campaign.

KERSHAW'S BRIGADE VOLUME 1 *by D. Augustus Dickert*—Manassas, Seven Pines, Sharpsburg (Antietam), Fredricksburg, Chancellorsville, Gettysburg, Chickamauga, Chattanooga, Fort Sanders & Bean Station.

KERSHAW'S BRIGADE VOLUME 2 *by D. Augustus Dickert*—At the wilderness, Cold Harbour, Petersburg, The Shenandoah Valley and Cedar Creek..

AVAILABLE ONLINE AT www.leonaur.com
AND FROM ALL GOOD BOOK STORES

07/09

LEONAUR

ALSO FROM LEONAUR
AVAILABLE IN SOFTCOVER OR HARDCOVER WITH DUST JACKET

THE RELUCTANT REBEL by William G. Stevenson—A young Kentuckian's experiences in the Confederate Infantry & Cavalry during the American Civil War..

BOOTS AND SADDLES by Elizabeth B. Custer—The experiences of General Custer's Wife on the Western Plains.

FANNIE BEERS' CIVIL WAR by Fannie A. Beers—A Confederate Lady's Experiences of Nursing During the Campaigns & Battles of the American Civil War.

LADY SALE'S AFGHANISTAN by Florentia Sale—An Indomitable Victorian Lady's Account of the Retreat from Kabul During the First Afghan War.

THE TWO WARS OF MRS DUBERLY by Frances Isabella Duberly—An Intrepid Victorian Lady's Experience of the Crimea and Indian Mutiny.

THE REBELLIOUS DUCHESS by Paul F. S. Dermoncourt—The Adventures of the Duchess of Berri and Her Attempt to Overthrow French Monarchy.

LADIES OF WATERLOO by Charlotte A. Eaton, Magdalene de Lancey & Juana Smith—The Experiences of Three Women During the Campaign of 1815: Waterloo Days by Charlotte A. Eaton, A Week at Waterloo by Magdalene de Lancey & Juana's Story by Juana Smith.

TWO YEARS BEFORE THE MAST by Richard Henry Dana. Jr.—The account of one young man's experiences serving on board a sailing brig—the Penelope—bound for California, between the years1834-36.

A SAILOR OF KING GEORGE by Frederick Hoffman—From Midshipman to Captain—Recollections of War at Sea in the Napoleonic Age 1793-1815.

LORDS OF THE SEA by A. T. Mahan—Great Captains of the Royal Navy During the Age of Sail.

COGGESHALL'S VOYAGES: VOLUME 1 by George Coggeshall—The Recollections of an American Schooner Captain.

COGGESHALL'S VOYAGES: VOLUME 2 by George Coggeshall—The Recollections of an American Schooner Captain.

TWILIGHT OF EMPIRE by Sir Thomas Ussher & Sir George Cockburn—Two accounts of Napoleon's Journeys in Exile to Elba and St. Helena: Narrative of Events by Sir Thomas Ussher & Napoleon's Last Voyage: Extract of a diary by Sir George Cockburn.

AVAILABLE ONLINE AT www.leonaur.com
AND FROM ALL GOOD BOOK STORES

07/09

![LEONAUR]

ALSO FROM LEONAUR
AVAILABLE IN SOFTCOVER OR HARDCOVER WITH DUST JACKET

ESCAPE FROM THE FRENCH *by Edward Boys*—A Young Royal Navy Midshipman's Adventures During the Napoleonic War.

THE VOYAGE OF H.M.S. PANDORA *by Edward Edwards R. N. & George Hamilton, edited by Basil Thomson*—In Pursuit of the Mutineers of the Bounty in the South Seas—1790-1791.

MEDUSA *by J. B. Henry Savigny and Alexander Correard and Charlotte-Adélaïde Dard* —Narrative of a Voyage to Senegal in 1816 & The Sufferings of the Picard Family After the Shipwreck of the Medusa.

THE SEA WAR OF 1812 VOLUME 1 *by A. T. Mahan*—A History of the Maritime Conflict.

THE SEA WAR OF 1812 VOLUME 2 *by A. T. Mahan*—A History of the Maritime Conflict.

WETHERELL OF H. M. S. HUSSAR *by John Wetherell*—The Recollections of an Ordinary Seaman of the Royal Navy During the Napoleonic Wars.

THE NAVAL BRIGADE IN NATAL *by C. R. N. Burne*—With the Guns of H. M. S. Terrible & H. M. S. Tartar during the Boer War 1899-1900.

THE VOYAGE OF H. M. S. BOUNTY *by William Bligh*—The True Story of an 18th Century Voyage of Exploration and Mutiny.

SHIPWRECK! *by William Gilly*—The Royal Navy's Disasters at Sea 1793-1849.

KING'S CUTTERS AND SMUGGLERS: 1700-1855 *by E. Keble Chatterton*—A unique period of maritime history-from the beginning of the eighteenth to the middle of the nineteenth century when British seamen risked all to smuggle valuable goods from wool to tea and spirits from and to the Continent.

CONFEDERATE BLOCKADE RUNNER *by John Wilkinson*—The Personal Recollections of an Officer of the Confederate Navy.

NAVAL BATTLES OF THE NAPOLEONIC WARS *by W. H. Fitchett*—Cape St. Vincent, the Nile, Cadiz, Copenhagen, Trafalgar & Others.

PRISONERS OF THE RED DESERT *by R. S. Gwatkin-Williams*—The Adventures of the Crew of the Tara During the First World War.

U-BOAT WAR 1914-1918 *by James B. Connolly/Karl von Schenk*—Two Contrasting Accounts from Both Sides of the Conflict at Sea During the Great War.

AVAILABLE ONLINE AT www.leonaur.com
AND FROM ALL GOOD BOOK STORES

07/09

LEONAUR

ALSO FROM LEONAUR
AVAILABLE IN SOFTCOVER OR HARDCOVER WITH DUST JACKET

IRON TIMES WITH THE GUARDS *by An O. E. (G. P. A. Fildes)*—The Experiences of an Officer of the Coldstream Guards on the Western Front During the First World War.

THE GREAT WAR IN THE MIDDLE EAST: 1 *by W. T. Massey*—The Desert Campaigns & How Jerusalem Was Won---two classic accounts in one volume.

THE GREAT WAR IN THE MIDDLE EAST: 2 *by W. T. Massey*—Allenby's Final Triumph.

SMITH-DORRIEN *by Horace Smith-Dorrien*—Isandlwhana to the Great War.

1914 *by Sir John French*—The Early Campaigns of the Great War by the British Commander.

GRENADIER *by E. R. M. Fryer*—The Recollections of an Officer of the Grenadier Guards throughout the Great War on the Western Front.

BATTLE, CAPTURE & ESCAPE *by George Pearson*—The Experiences of a Canadian Light Infantryman During the Great War.

DIGGERS AT WAR *by R. Hugh Knyvett & G. P. Cuttriss*—"Over There" With the Australians by R. Hugh Knyvett and Over the Top With the Third Australian Division by G. P. Cuttriss. Accounts of Australians During the Great War in the Middle East, at Gallipoli and on the Western Front.

HEAVY FIGHTING BEFORE US *by George Brenton Laurie*—The Letters of an Officer of the Royal Irish Rifles on the Western Front During the Great War.

THE CAMELIERS *by Oliver Hogue*—A Classic Account of the Australians of the Imperial Camel Corps During the First World War in the Middle East.

RED DUST *by Donald Black*—A Classic Account of Australian Light Horsemen in Palestine During the First World War.

THE LEAN, BROWN MEN *by Angus Buchanan*—Experiences in East Africa During the Great War with the 25th Royal Fusiliers—the Legion of Frontiersmen.

THE NIGERIAN REGIMENT IN EAST AFRICA *by W. D. Downes*—On Campaign During the Great War 1916-1918.

THE 'DIE-HARDS' IN SIBERIA *by John Ward*—With the Middlesex Regiment Against the Bolsheviks 1918-19.

AVAILABLE ONLINE AT **www.leonaur.com**
AND FROM ALL GOOD BOOK STORES

07/09

LEONAUR

ALSO FROM LEONAUR

AVAILABLE IN SOFTCOVER OR HARDCOVER WITH DUST JACKET

FARAWAY CAMPAIGN *by F. James*—Experiences of an Indian Army Cavalry Officer in Persia & Russia During the Great War.

REVOLT IN THE DESERT *by T. E. Lawrence*—An account of the experiences of one remarkable British officer's war from his own perspective.

MACHINE-GUN SQUADRON *by A. M. G.*—The 20th Machine Gunners from British Yeomanry Regiments in the Middle East Campaign of the First World War.

A GUNNER'S CRUSADE *by Antony Bluett*—The Campaign in the Desert, Palestine & Syria as Experienced by the Honourable Artillery Company During the Great War .

DESPATCH RIDER *by W. H. L. Watson*—The Experiences of a British Army Motorcycle Despatch Rider During the Opening Battles of the Great War in Europe.

TIGERS ALONG THE TIGRIS *by E. J. Thompson*—The Leicestershire Regiment in Mesopotamia During the First World War.

HEARTS & DRAGONS *by Charles R. M. F. Crutwell*—The 4th Royal Berkshire Regiment in France and Italy During the Great War, 1914-1918.

INFANTRY BRIGADE: 1914 *by John Ward*—The Diary of a Commander of the 15th Infantry Brigade, 5th Division, British Army, During the Retreat from Mons.

DOING OUR 'BIT' *by Ian Hay*—Two Classic Accounts of the Men of Kitchener's 'New Army' During the Great War including *The First 100,000 & All In It.*

AN EYE IN THE STORM *by Arthur Ruhl*—An American War Correspondent's Experiences of the First World War from the Western Front to Gallipoli-and Beyond.

STAND & FALL *by Joe Cassells*—With the Middlesex Regiment Against the Bolsheviks 1918-19.

RIFLEMAN MACGILL'S WAR *by Patrick MacGill*—A Soldier of the London Irish During the Great War in Europe including *The Amateur Army, The Red Horizon & The Great Push.*

WITH THE GUNS *by C. A. Rose & Hugh Dalton*—Two First Hand Accounts of British Gunners at War in Europe During World War 1- Three Years in France with the Guns and With the British Guns in Italy.

THE BUSH WAR DOCTOR *by Robert V. Dolbey*—The Experiences of a British Army Doctor During the East African Campaign of the First World War.

AVAILABLE ONLINE AT www.leonaur.com
AND FROM ALL GOOD BOOK STORES

07/09

ALSO FROM LEONAUR

AVAILABLE IN SOFTCOVER OR HARDCOVER WITH DUST JACKET

THE 9TH—THE KING'S (LIVERPOOL REGIMENT) IN THE GREAT WAR 1914 - 1918 *by Enos H. G. Roberts*—Mersey to mud—war and Liverpool men.

THE GAMBARDIER *by Mark Severn*—The experiences of a battery of Heavy artillery on the Western Front during the First World War.

FROM MESSINES TO THIRD YPRES *by Thomas Floyd*—A personal account of the First World War on the Western front by a 2/5th Lancashire Fusilier.

THE IRISH GUARDS IN THE GREAT WAR - VOLUME 1 *by Rudyard Kipling*—Edited and Compiled from Their Diaries and Papers—The First Battalion.

THE IRISH GUARDS IN THE GREAT WAR - VOLUME 1 *by Rudyard Kipling*—Edited and Compiled from Their Diaries and Papers—The Second Battalion.

ARMOURED CARS IN EDEN *by K. Roosevelt*—An American President's son serving in Rolls Royce armoured cars with the British in Mesopatamia & with the American Artillery in France during the First World War.

CHASSEUR OF 1914 *by Marcel Dupont*—Experiences of the twilight of the French Light Cavalry by a young officer during the early battles of the great war in Europe.

TROOP HORSE & TRENCH *by R.A. Lloyd*—The experiences of a British Lifeguardsman of the household cavalry fighting on the western front during the First World War 1914-18.

THE EAST AFRICAN MOUNTED RIFLES *by C.J. Wilson*—Experiences of the campaign in the East African bush during the First World War.

THE LONG PATROL *by George Berrie*—A Novel of Light Horsemen from Gallipoli to the Palestine campaign of the First World War.

THE FIGHTING CAMELIERS *by Frank Reid*—The exploits of the Imperial Camel Corps in the desert and Palestine campaigns of the First World War.

STEEL CHARIOTS IN THE DESERT *by S. C. Rolls*—The first world war experiences of a Rolls Royce armoured car driver with the Duke of Westminster in Libya and in Arabia with T.E. Lawrence.

WITH THE IMPERIAL CAMEL CORPS IN THE GREAT WAR *by Geoffrey Inchbald*—The story of a serving officer with the British 2nd battalion against the Senussi and during the Palestine campaign.

AVAILABLE ONLINE AT www.leonaur.com
AND FROM ALL GOOD BOOK STORES

07/09

www.ingramcontent.com/pod-product-compliance
Lightning Source LLC
Chambersburg PA
CBHW032043080426
42733CB00006B/175